This book presents an expansion of the highly successful lectures given by Professor Ladyzhenskaya at the University of Rome, 'La Sapienza', under the auspices of the Accademia dei Lincei. The lectures were devoted to questions of the behaviour of trajectories for semi-groups of nonlinear bounded continuous operators in a locally non-compact metric space and for solutions of abstract evolution equations. The latter contain many boundary value problems for partial differential equations of a dissipative type. Professor Ladyzhenskaya is an internationally renow     nematician and her lectures attracted large audiences. These notes reflec          alibre of her lectures and should prove essential reading for a          .crested in partial differential equations and dynamical systems.

DUE FOR RETURN

*Attractors for semigroups
and evolution equations*

Lezioni Lincee
*Editor: Luigi A. Radicati di Brozolo, Scuola Normale Superiore,
Pisa*

This series of books arises from lectures given under the auspices
of the Accademia Nazionale dei Lincee through a grant from IBM
Italia.
The lectures, given by international authorities, will range on
scientific topics from mathematics and physics through to biology
and economics. The books are intended for a broad audience of
graduate students and faculty members, and are meant to provide
a '*mise au point*' for the subject they deal with.
The symbol of the Accademia, the Lynx, is noted for its sharp
sightedness; the volumes in the series will be penetrating studies
of scientific topics of contemporary interest.

*Already published*

# Attractors for semigroups and evolution equations

## OLGA LADYZHENSKAYA

*Mathematical Institute of Academy USSR, Leningrad*

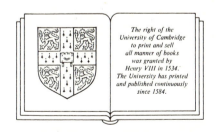

The right of the
University of Cambridge
to print and sell
all manner of books
was granted by
Henry VIII in 1534.
The University has printed
and published continuously
since 1584.

## CAMBRIDGE UNIVERSITY PRESS

CAMBRIDGE
NEW YORK PORT CHESTER
MELBOURNE SYDNEY

Published by the Press Syndicate of the University of Cambridge
The Pitt Building, Trumpington Street, Cambridge, CB2 1RP
40 West 20th Street, New York, NY 10011-4211, USA
10 Stamford Road, Oakleigh, Melbourne 3166, Australia

First published 1991

Printed in Great Britain at the University Press, Cambridge

A catalogue record for this book is available from the British Library

Library of Congress cataloguing in publication data available


ISBN 0 521 39030 3  hardback
ISBN 0 521 39922 X  paperback

# Contents

# Preface

These lecture notes are devoted to questions of the behaviour, when $t \to \infty$, of trajectories $V_t(v)$, $t \in \mathbf{R}^+ = [0, \infty)$ for semigroups $\{V_t, t \in \mathbf{R}^+, X\}$ of nonlinear bounded continuous operators $V_t$ in a locally non-compact metric space $X$ and for solutions of abstract evolution equations. The latter contain many boundary value problems for PDE (partial differential equations) of a dissipative type.

In contrast to the traditional theory of the local stability of PDE (i.e. in the vicinity of a solution) we study the behaviour of all trajectories or solutions for the problems and give a description of the set of all limit states. We will not make assumptions either about the smallness of the parameters in the problem or on the closeness of the problem to a linear one, neither will we consider any other condition that ensures that all the solutions of the problem tend to some special solution. Our purpose is to develop a global theory of stability for problems of mathematical physics with dissipation. The principal ideas in this subject were formulated in paper [1] and I follow them here. The object of paper [1] concerns boundary value problems for Navier–Stokes equations. This object helped us to understand which properties of semigroup $\{V_t, t \in \mathbf{R}^+, X\}$ imply the compactness of the set of all limit states (or, which is the same, the minimal global $B$-attractor), its invariance, the possibility of continuing the semigroup restricted on $\mathcal{M}$ to the full group on $\mathcal{M}$ and a finiteness of dynamics $\{V_t, t \in \mathbf{R} = (-\infty, +\infty), \mathcal{M}\}$.

The latter was a source of investigations of the finiteness of dimensions of compact sets in the Hilbert space $X$ which are invariant under a nonlinear bounded operator $V$ enjoying some special properties. The paper by Mallet-Paret [2] was the first one in this direction. In the paper by Douady and Oesterlé [3], the Hausdorff dimension ($\dim_H \mathcal{A}$) of such sets $\mathcal{A}$ was estimated for a wider class of operators $V$. Although the full proof in [3] was done for the case of a Euclidean space $X \equiv \mathbf{R}^n$, the authors pointed out that it may be generalized to the case of a Hilbert space $X$.

After these papers many works devoted to such question were published ([4]–[10], etc.). In the first part of Chapter 4 we evaluate Hausdorff and fractal dimensions of compact invariant sets following the approach of paper [3] (see [11]). In the second part of Chapter 4 we show how to verify the conditions of our theorems in the case of semigroups generated by evolution equations.

Let us mention that in paper [6] there is a theorem with a very short and clear proof which was used for estimating both $\dim_H \mathcal{A}$ and the fractal dimension for many PDE of different types. But majorants obtained through this theorem are worse than majorants deduced from theorems of Chapter 4 (see Part II of these lectures and [11]).

We do not give here the full list of papers relating to attractors of PDE. In the eighties several papers on this subject have been published, and the number continually increases. A survey of the relevant papers published before 1986 can be found in [12]. I would like to point out that there are many connections between the results of Chapters 2 and 3 and the results of American mathematicians from Brown University. The latter were developed in the process of studying ODE (ordinary differential equations) with delay and abstract discrete semigroups. They have been expounded in the monograph [13] by J.K. Hale. Semi-linear parabolic equations (mostly with one space argument) are considered in the book [14] by D. Henry, who is concerned only with the investigations of American mathematicians and does not seem to be aware of paper [1]; actually in the first lines of [14] he expresses the wish that attractors of the Navier–Stokes equations and some other problems of hydrodynamics for viscous fluids be investigated.

I would like to express my cordial thanks to members of the Academia Nazionale dei Lincei and to Professor G. Fichera especially for the invitation to deliver these lectures and to publish them. I am very much obliged to Professors G. Fichera, P. Castellani and M. Sneider for their help in the preparation of the English version of my lectures.

# Part I.
# Attractors
# for the semigroups
# of operators:
# an abstract
# framework

# 1

---

# Basic notions

In this chapter we shall deal with semigroups $\{V_t, t \in \mathbf{R}^+ = [0, +\infty)\}$ of continuous operators $V_t : X \to X$ acting on a complete metric space $X$. We shall denote them $\{V_t, t \in \mathbf{R}^+, X\}$ or simply $\{V_t\}$.

In what follows, the term *semigroup* refers to any family of single-valued continuous operators $V_t : X \to X$ depending on a parameter $t \in \mathbf{R}^+$ and enjoying the semigroup property: $V_{t_1}(V_{t_2}(x)) = V_{t_1+t_2}(x)$ for all $t_1, t_2 \in \mathbf{R}^+$ and $x \in X$.

A semigroup $\{V_t\}$ is called *pointwise continuous* if the mapping $t \to V_t(x)$ from $\mathbf{R}^+$ to $X$ is continuous for each $x \in X$. A semigroup is called *continuous* if the mapping $(t, x) \to V_t(x)$ from $\mathbf{R}^+ \times X$ to $X$ is continuous.

Given a semigroup $\{V_t\}$ the following notation will be frequently used:

$$\gamma^+(x) := \{y \in X \mid y = V_t(x), \, t \in \mathbf{R}^+\}$$
$$\equiv \{V_t(x), \, t \in \mathbf{R}^+\} \, ;$$
$$\gamma^+_{[t_1,t_2]}(x) := \{V_t(x), \, t \in [t_1, t_2]\} \, ;$$
$$\gamma^+_t(x) := \gamma^+_{[t,\infty)}(x) \equiv \{V_\tau(x), \, \tau \in [t, \infty)\} \, ;$$
$$\gamma^+(A) := \bigcup_{x \in A} \gamma^+(x) \, ;$$
$$\gamma^+_{[t_1,t_2]}(A) := \bigcup_{x \in A} \gamma^+_{[t_1,t_2]}(x) \, ;$$
$$\gamma^+_t(A) := \bigcup_{x \in A} \gamma^+_t(x) \, .$$

It is easy to verify that $V_t(\gamma^+(A)) = \gamma_t^+(A)$.

The curve $\gamma^+(x)$ is called the positive semi-trajectory of $x$.

The collection of all bounded subsets of $X$ is denoted by $\mathcal{B}$.

We use the letter $B$ (with or without indices) to denote the elements of $\mathcal{B}$, i.e. the bounded subsets of $X$.

A semigroup $\{V_t\}$ is called *locally bounded* if $\gamma_{[0,t]}^+(B) \in \mathcal{B}$ for all $B \in \mathcal{B}$ and all $t \in \mathbb{R}^+$. $\{V_t\}$ is a *bounded semigroup* if $\gamma^+(B) \in \mathcal{B}$ for each $B \in \mathcal{B}$.

Let $A$ and $M$ be subsets of $X$. We say that $A$ *attracts* $M$ or $M$ *is attracted* to $A$ by semigroup $\{V_t\}$ if for every $\epsilon > 0$ there exists a $t_1(\epsilon, M) \in \mathbb{R}^+$ such that $V_t(M) \subset \mathcal{O}_\epsilon(A)$ for all $t \geq t_1(\epsilon, M)$. Here $\mathcal{O}_\epsilon(A)$ is the $\epsilon$-neighbourhood of $A$ (i.e. the union of all open balls of radii $\epsilon$ centered at the points of $A$). We say that the set $A \subset X$ *attracts the point* $x \in X$ if $A$ attracts the one-point set $\{x\}$.

If $A$ attracts each point $x$ of $X$ then $A$ is called a *global attractor* (for the semigroup). $A$ is called a *global $B$-attractor* if $A$ attracts each bounded set $B \in \mathcal{B}$.

A semigroup is called *pointwise dissipative* (respectively, *$B$-dissipative*) if it has a bounded global attractor (respectively a bounded global $B$-attractor).

Our main purpose here is to find those semigroups for which there is a *minimal closed global $B$-attractor* and investigate properties of such attractors. These attractors will be designated by $\mathcal{M}$. We shall examine also the existence of a *minimal closed global attractor* $\widehat{\mathcal{M}}$. It is clear that $\widehat{\mathcal{M}} \subset \mathcal{M}$ and later on we will also verify that $\widehat{\mathcal{M}}$ might be just a small part of $\mathcal{M}$.

The concept of invariant sets is closely related to these subjects. We call a set $A \subset X$ *invariant* (relative to semigroup $\{V_t\}$) if $V_t(A) = A$ for all $t \in \mathbb{R}^+$.

A set $A \subset X$ is called *absorbing* if for every $x \in X$ there exists a $t_1(x) \in \mathbb{R}^+$ such that $V_t(x) \in A$ for all $t \geq t_1(x)$. A set $A$ is called *$B$-absorbing* if for every $B \in \mathcal{B}$ there exists a $t_1(B) \in \mathbb{R}^+$ such that $V_t(B) \subset A$ for all $t \geq t_1(B)$.

In our investigation of the problems concerning the attractors $\mathcal{M}$ and $\widehat{\mathcal{M}}$ the concept of $\omega$-limit sets will play a fundamental role. For $x \in X$ the *$\omega$-limit set* $\omega(x)$ is, by definition, the set of all $y \in X$ such that $y = \lim_{k \to \infty} V_{t_k}(x)$ for a sequence $t_k \nearrow +\infty$.

The *$\omega$-limit set* $\omega(A)$ for a set $A \subset X$ is the set of the limits of all converging sequences of the form $V_{t_k}(x_k)$, where $x_k \in A$ and $t_k \nearrow +\infty$.

An equivalent description of the $\omega$-limit sets is given by

**Lemma 1.1**

$$\omega(x) = \bigcap_{t \geq 0} [\gamma_t^+(x)]_X \; ; \quad \omega(A) = \bigcap_{t \geq 0} [\gamma_t^+(A)]_X \; . \tag{1.1}$$

Here the symbol $[\;\;]_X$ means the closure in the topology of the metric space $X$.

The proof of Lemma 1.1 is traditional and so is omitted. Since $\gamma_{t_2}^+(A) \subset \gamma_{t_1}^+(A)$ whenever $t_2 > t_1$, the intersection over all $t \in \mathbf{R}^+$ in (1.1) may be replaced by $\bigcap_{t \geq T}$ with any $T \in \mathbf{R}^+$.

It is necessary to have in mind that for locally non-compact spaces $X$ the use of the concept of limit sets requires some caution since the intersection $A_0 = \bigcap_{k=1}^{\infty} A_k$ of $A_k = [A_k]_X \supset A_{k+1} = [A_{k+1}]_X$ in them may be empty (and therefore unhelpful).

# 2

---

# Semigroup of class $\mathcal{K}$

By definition a semigroup $\{V_t\}$ belongs to the *class* $\mathcal{K}$ if for each $t > 0$ the operator $V_t$ is compact, i.e. for any bounded set $B \subset X$ its image $V_t(B)$ is precompact. (I should remind readers that the operators $V_t$ are supposed to be continuous, see the beginning of Chapter 1).

### Theorem 2.1

Let the semigroup $\{V_t\}$ belong to the class $\mathcal{K}$. Let $A \subset X$ and $T \in \mathbf{R}^+$. Suppose that $\gamma_T^+(A) \in \mathcal{B}$. Then

(i) $\omega(A)$ is non-empty and compact,

(ii) $\omega(A)$ attracts $A$,

(iii) $\omega(A)$ is invariant, i.e. $V_t(\omega(A)) = \omega(A)$ for all $t \in \mathbf{R}^+$,

(iv) $\omega(A)$ is the minimal closed set which attracts $A$,

(v) $\omega(A)$ is connected provided $A$ is connected and the semigroup $\{V_t\}$ is continuous.

*Proof* The sets $V_t(\gamma_T^+(A)) = \gamma_{t+T}^+(A)$, $0 < t < +\infty$, are precompact and $\gamma_{t_2+T}^+(A) \subset \gamma_{t_1+T}^+(A)$ for all $t_2 > T_1$. Therefore $\omega(A) = \bigcap_{t>0} [\gamma_{t+T}^+(A)]_X$ is the intersection of an ordered family of compact sets. Hence $\omega(A)$ is non-empty, compact and attracts $A$.

In order to prove (iii) we verify immediately that $V_t(\omega(A)) \subset \omega(A)$. Actually, if $y \in \omega(A)$ then $y = \lim_{k\to\infty} V_{t_k}(x_k)$ for some $x_k \in A$ and $t_k \nearrow +\infty$, hence $V_t(y) = V_t(\lim_{k\to\infty} V_{t_k}(x_k)) = \lim_{k\to\infty} V_{t+t_k}(x_k)$ and thus $V_t(y) \in \omega(A)$.

To obtain the reverse inclusion, $\omega(A) \subset V_t(\omega(A))$, let $x \in \omega(A)$

and therefore $x = \lim_{k \to \infty} V_{t_k}(x_k)$ for some $x_k \in A$ and $t_k \nearrow +\infty$. We may assume that $1 + T + t \leq t_1 < t_2 < \ldots$. The points $y_k = v_{t_k - t}(x_k)$, $k = 1, 2, \ldots$, belong to the precompact set $\gamma_{T+1}^+(A)$. Hence there is a converging subsequence $\{y_{k_j}\}$ and $\lim_{j \to \infty} y_{k_j} = y \in \omega(A)$.

Consequently $x = \lim_{j \to \infty} V_{t_{k_j}}(x_{k_j}) = \lim_{j \to \infty} V_t(y_{k_j}) = V_t(y)$. Thus we obtain $\omega(A) \subset V_t(\omega(A))$ and (iii) is proved.

To prove the minimality of $\omega(A)$ suppose the contrary and let $F$ be a proper closed subset of $\omega(A)$ which attracts $A$. As $\omega(A)$ is compact so is $F$. Choose any $y \in \omega(A) \setminus F$. For $\epsilon > 0$ small enough the $\epsilon$-neighbourhoods $\mathcal{O}_\epsilon(y)$ and $\mathcal{O}_\epsilon(F)$ do not intersect. We assumed that $F$ attracts $A$. Hence $V_t(A) \subset \mathcal{O}_\epsilon(F)$ for all $t \geq t(\epsilon)$ with some $t(\epsilon) \geq 0$. On the other hand $y = \lim_{k \to \infty} V_{t_k}(x_k)$ for some $x_k \in A$ and $t_k \nearrow +\infty$ (since $y \in \omega(A)$). Consequently, $V_{t_k}(A) \cap \mathcal{O}_\epsilon(y) \neq \emptyset$ for $t_k$ large enough. Hence $\mathcal{O}_\epsilon(F) \cap \mathcal{O}_\epsilon(y) \neq \emptyset$, a contradiction.

Now let $A$ be connected and the semigroup be continuous. Suppose that $\omega(A)$ is not connected. Then we may decompose $\omega(A)$ as follows: $\omega(A) = F_1 \cup F_2$, where $F_1$ and $F_2$ are non-empty closed disjoint sets. Therefore open $\epsilon$-neighbourhoods $\mathcal{O}_\epsilon(F_1)$ and $\mathcal{O}_\epsilon(F_2)$ do not intersect for $\epsilon > 0$ small enough. We have $\mathcal{O}_\epsilon(\omega(A)) = \mathcal{O}_\epsilon(F_1) \cup \mathcal{O}_\epsilon(F_2)$.

Since $\omega(A)$ attracts $A$, there is some $t_1 \equiv t_1(\epsilon, A)$ such that $\gamma_t^+(A) \subset \mathcal{O}_\epsilon(\omega(A))$ for all $t \geq t_1$. But $\gamma_t^+(A)$ is connected since it is the continuous image of $[t, +\infty) \times A$ (under the mapping $(\tau, x) \to V_\tau(x)$). So for all $t \geq t_1$ either $\gamma_t^+(A) \subset \mathcal{O}_\epsilon(F_1)$ or $\gamma_t^+(A) \subset \mathcal{O}_\epsilon(F_2)$. Consequently, either $\omega(A) \subset F_1$ or $\omega(A) \subset F_2$, hence either $F_1 = \emptyset$ or $F_2 = \emptyset$; this is a contradiction. Thus, $\omega(A)$ must be connected. ∎

By definition, a *complete trajectory* $\gamma(x)$ *of the point* $x$ is the curve $x(t)$, $-\infty < t < +\infty$, satisfying the following conditions: $x(t) \in X$ for all $t \in \mathbf{R}$, $x(0) = x$, $V_\tau(x(t)) = x(t + \tau)$ for all $t \in \mathbf{R}$ and $\tau \in \mathbf{R}^+$. The set $\gamma^-(x) = \{x(t), -\infty < t \leq 0\}$ is called a *negative semi-trajectory of* $x$. Thus, $\gamma(x) = \gamma^+(x) \cup \gamma^-(x)$. In general, for an arbitrary $x \in X$, a complete trajectory $\gamma(x)$ may not exist and even if it does, it might be not unique. However the following statement is true.

### Lemma 2.1

*Let $A$ be an invariant set (i.e., $V_t(A) = A$ for all $t \in \mathbf{R}^+$). Then for every $x \in A$ there exists a complete trajectory $\gamma(x)$. If the semigroup $\{V_t\}$ is pointwise continuous then the trajectory $\gamma(x)$ is a* *continuous curve in A. If the operators $V_t$, $t \in \mathbf{R}^+$, are invertible on A then:*

(i) *through each $x \in A$ passes a unique trajectory $\gamma(x)$;*

(ii) *the family $\{V_t, t \in \mathbf{R}, A\}$, where $V_t := V_{-t}^{-1}$ for $t < 0$, has the group property: $V_{t_1+t_2} = V_{t_1}V_{t_2}$ for any $t_1, t_2 \in \mathbf{R}$; if additionally A is compact, then $\{V_t, t \in \mathbf{R}, A\}$ is the group of continuous operators. This group is pointwise continuous or continuous if $\{V_t, t \in \mathbf{R}^+, A\}$ is pointwise continuous or continuous correspondingly.*

We omit the proof of the lemma since it is traditional. Let us describe only the construction of $\gamma(x)$, $x \in A$. For $x \in A$ there is at least one point $x_{-1} \in A$ for which $V_1(x_{-1}) = x$; for $x_{-1}$ there is a point $x_{-2} \in A$ for which $V_1(x_{-2}) = x_{-1}$ and so on. Let us join the points $x_{-k-1}$ and $x_{-k}$ by the curve $\{V_t(x_{-k-1}), t \in [0,1]\}$. The collection of all these curves for all $k = 0, 1, \ldots$ forms $\gamma^-(x)$ and $x(t) = V_{t+k+1}(x_{-k-1})$, for $t \in [-k-1, -k]$.

Now we turn to the problem of the existence of the minimal global $B$-attractor $\mathcal{M}$ for a semigroup $\{V_t\}$ of class $\mathcal{K}$.

Consider first the simplest case when there exists a global $B$-absorbing bounded set $B_0 \in \mathcal{B}$. Then for every $B \in \mathcal{B}$ there is $T(B) \geq 0$ so that $V_t(B) \subset B_0$ for all $t \geq T(B)$. In particular, $V_t(B_0) \subset B_0$ for all $t \geq T(B_0)$ and consequently $\gamma^+_{T(B_0)}(B_0) \subset B_0$. In view of Theorem 2.1 the set $\omega(B_0)$ is a non-empty compact invariant set. Moreover, $\omega(B_0)$ attracts $B_0$. Hence, for every $\epsilon > 0$ there exists $t_1(\epsilon) \geq 0$ such that $V_t(B_0 \subset \mathcal{O}_\epsilon(\omega(B_0))$ for all $t \geq t_1(\epsilon)$. Therefore, given any $B \in \mathcal{B}$ we have $V_t(B) \subset \mathcal{O}_\epsilon(\omega(B_0))$ for all $t \geq t_1(\epsilon) + T(B)$. Hence, $\omega(B_0)$ is a closed global $B$-attractor. It is minimal due to Theorem 2.1 (iv). Thus, $\omega(B_0) = \mathcal{M}$.

Assume now that the semigroup $\{V_t\}$ is $B$-dissipative and $B_1$ is its bounded global $B$-attractor. Then $\mathcal{O}_{\epsilon_1}(B_1)$ (with $\epsilon_1 > 0$) is a global $B$-absorbing bounded set, and due to the first case $\mathcal{M} = \omega(\mathcal{O}_{\epsilon_1}(B_1))$.

Consider now a more complicated situation. Suppose that $\{V_t\}$ is a bounded and pointwise dissipative semigroup. In particular, there is a bounded global attractor, say $B_2$. Choose $\epsilon_2 > 0$, and put $B_1 := \mathcal{O}_{\epsilon_2}(B_2)$ and $B_0 := \gamma^+(B_1)$. We are going to prove that $\omega(B_1) = \mathcal{M}$ is the minimal global $B$-attractor.

Indeed, since $B_2$ is a global attractor, for every point $x \in X$ there is $T(x) \geq 0$ such that $V_{T(x)}(x) \in B_1 = \mathcal{O}_{\epsilon_2}(B_2)$. Since $B_1$ is an open set and the operator $V_{T(x)}$ is continuous, we have $V_{T(x)}(\mathcal{O}_{\epsilon(x)}(x)) \subset B_1$, for some $\epsilon(x) > 0$. Hence $V_{t+T(x)}(\mathcal{O}_{\epsilon(x)}(x)) \subset V_t(B_1) \subset \gamma^+(B_1) = B_0$ for all $t \geq 0$. Now, by standard arguments, for every compact set $K$ there are $\epsilon(K) > 0$ and $T(K) \geq 0$ such that

$$V_t(\mathcal{O}_{\epsilon(K)}(K)) \subset B_0 \qquad \text{for all} \quad t \geq T(K). \tag{2.1}$$

In view of Theorem 2.1 every bounded set $B$ is attracted to its $\omega$-limit set $\omega(B)$. Hence, $V_t(B) \subset \mathcal{O}_{\epsilon_1}(\omega(B))$ for all $t \geq t_1(\epsilon_1, B)$. Since $\omega(B)$ is compact we may choose $\epsilon_1 = \epsilon(\omega(B))$ and deduce from (2.1) that $V_{t+t_1}(B) \subset B_0$ for all $t \geq T(\omega(B))$ and $t_1 = t_1(\epsilon_1, B)$. Thus, $B_0$ is a global $B$-absorbing bounded set.

Hence, as was shown above, $\omega(B_0) = \mathcal{M}$. Note that $V_t(B_0) = \gamma_t^+(B_1)$ by definition of $B_1$ and $B_0$, and $\gamma_t^+(B_1) \to \omega(B_1)$ when $t \to \infty$, therefore $\omega(B_0) = \omega(B_1)$. For minimality of $\omega(B_1)$ see Theorem 2.1 (iv).

If there is a connected $B \supset \mathcal{M}$ then $\mathcal{M}$ is connected since $V_t(B)$, $t \in \mathbf{R}^+$ are connected and for any $\epsilon > 0$, $\mathcal{M} = V_t(\mathcal{M}) \subset V_t(B) \subset \mathcal{O}_\epsilon(\mathcal{M})$ for $t \geq t_1(\epsilon, B)$.

Thus we have proved the following

### Theorem 2.2

*Let $\{V_t, t \in \mathbf{R}^+, X\}$ be a semigroup of class $\mathcal{K}$. Suppose that it is either B-dissipative or bounded and pointwise dissipative. Then $\{V_t, t \in \mathbf{R}^+, X\}$ has a minimal global B-attractor $\mathcal{M}$, which is compact and invariant. $\mathcal{M}$ is connected provided so is $X$.*

As a by-product of the above considerations (see (2.1)) we have proved:

### Proposition 2.1

*If the semigroup $\{V_t, t \in \mathbf{R}^+, X\}$ is bounded and pointwise dissipative then there is a bounded set $B_0$ such that for every compact $K$ (2.1) holds with some $\epsilon(K) > 0$ and $T(K) \geq 0$ and $V_t(B_0) \subset B_0$ for all $t \in \mathbf{R}^+$.*

The next proposition provides useful information on the structure of the minimal global $B$-attractor $\mathcal{M}$.

### Proposition 2.2

*Under the assumptions of Theorem 2.2 the minimal global B-attractor $\mathcal{M}$ may be characterized as follows:*

(i) $\mathcal{M} = \bigcup_{B \in \mathcal{B}} \omega(B)$;

(ii) $\mathcal{M} = \bigcup_{K \in \mathcal{K}} \omega(K)$ where $\mathcal{K}$ is the collection of all compact sets in $X$;

(iii) $\mathcal{M}$ is the union of all complete bounded trajectories in $X$;

(iv) $\mathcal{M}$ is the union of all complete precompact trajectories in $X$;

(v) $\mathcal{M}$ is the maximal invariant bounded set in $X$;

(vi) $\widehat{\mathcal{M}} = [\bigcup_{x \in X} \omega(x)]_X$.

*Proof*

(i) Every bounded set $B$ is attracted to its $\omega$-limit set $\omega(B)$ and to $\mathcal{M}$ and, hence, to $\omega(B) \cap \mathcal{M}$. Since $\omega(B)$ is minimal, $\omega(B)$ must entirely lie in $\mathcal{M}$, i.e. $\omega(B) \subset \mathcal{M}$. Now, $\mathcal{M}$ being invariant and compact, we have $\omega(\mathcal{M}) \equiv \mathcal{M}$. Thus, (i) is proved.

(ii) Since $\mathcal{M}$ and $\omega(B)$ for each $B \in \mathcal{B}$, are invariant compact sets, (ii) follows from (i).

(iii) and (iv) In view of Lemma 2.2, through every point $x \in \mathcal{M}$ passes a complete trajectory $\gamma(x)$. Any such trajectory lies in $\mathcal{M}$ and thus is bounded (and precompact). On the other hand, let $\gamma(x) = \{x(t),\, t \in \mathbf{R}\}$ be a bounded complete trajectory passing through some point $x = x(0) \in X$. Since $\gamma(x)$ is invariant (and bounded), then $\gamma(x)$ is precompact. Hence, $B := [\gamma(x)]_X$ is a compact invariant set. Therefore $\omega(B) \equiv B$ and so $B \subset \mathcal{M}$ (see (i) above).

(v) If $B$ is a bounded invariant set, then $V_t(B) = B$ and therefore $\omega(B) = B$ and $B \subset \mathcal{M}$. On the other hand $\mathcal{M}$ is a bounded invariant set.

(vi) is obvious.

This concludes the proof of the proposition. ∎

The structure of $\mathcal{M}$ is simpler than in the general case, if for the semigroup $\{V_t,\, t \in \mathbf{R}^+,\, X\}$ there is a "good" ("strong") Lyapunov function, i.e. a continuous function $\mathcal{L}: X \to \mathbf{R}$ which strongly decreases along each $\gamma^+(x) : \mathcal{L}(V_t(x)) \searrow$ when $t \nearrow$ (except, of course, stationary points: $z = V_t(z)$). Let $Z$ be the set of all stationary points of $\{V_t\}$. If the semigroup $\{V_t\}$ belongs to the class $\mathcal{K}$ and $\gamma^+(x) \in \mathcal{B}$ for any $x \in X$, then $Z$ is its minimal global attractor $\widehat{\mathcal{M}}$. In fact, for any $x \in X$ there exists a $\lim_{t \to \infty} \mathcal{L}(V_t(x)) \equiv \ell_+(x)$, a compact

$\omega(x)$ and $\mathcal{L}_{|\omega(x)} \equiv \ell_+(x) =$ constant. Therefore $\omega(x) \subset Z$ and $x$ is attracted to $Z$. If $Z$ is a bounded set then the semigroup $\{V_t\}$ is pointwise dissipative and Theorem 2.2 guarantees the existence of a compact attractor $\mathcal{M}$ provided $\{V_t\}$ is bounded. For each $x \in \mathcal{M}$, we can take a complete trajectory $\gamma(x) = \{x(t),\ t \in \mathbf{R},\ x(0) = x\}$ lying in $\mathcal{M}$ and determine for it the $\alpha$-limit set $\alpha(\gamma(x)) := \bigcap_{r \le 0} [\gamma_r^-(x)]x$ where $\gamma_r^-(x) := \{x(t),\ t \le \tau\}$. This $\alpha$-limit, set like, $\omega(x)$ is non-empty, invariant and $x(t) \to \alpha(\gamma(x))$ when $t \to -\infty$. It also belongs to $Z$ since $\lim_{t \to -\infty} \mathcal{L}(x(t)) =$ constant $= \mathcal{L}_{|\alpha(\gamma(x))}$. So we may say that both ends of trajectory $\gamma(x)$ tend to $Z$. If for example, the space $X$ is connected and $Z$ is not connected, then $\mathcal{M}$ (which is connected) contains not only points of $Z$ but complete trajectories connecting points of $Z$ (so $Z$ is smaller than $\mathcal{M}$).

Let us summarize these facts:

**Theorem 2.3**

*Suppose that the semigroup $\{V_t,\ t \in \mathbf{R}^+,\ X\}$ belongs to the class $\mathcal{K}$ and $\gamma^+(x) \in \mathcal{B}$ for any $x \in X$. If for this semigroup there is a "good" Lyapunov function $\mathcal{L}$, then its minimal global attractor $\widehat{\mathcal{M}}$ is non-empty and coincides with the set $Z$ of all stationary points. If $Z$ is a bounded set and $\{V_t\}$ is bounded then the semigroup has a minimal global B-attractor $\mathcal{M}$ enjoying properties indicated in Theorem 2.2. Both ends of any complete trajectory $\gamma(x) \subset \mathcal{M}$ tend to $Z$ (when $t \to \pm\infty$ correspondingly). If $X$ is connected and $Z$ is not, then $Z$ is a proper part of $\mathcal{M}$ and the, attractor $\mathcal{M}$ consists of complete trajectories which connect points of $Z$.*

# 3

---

# Semigroups of class $A\mathcal{K}$

A semigroup $\{V_t, t \in \mathbf{R}^+, X\}$ belongs to the class $A\mathcal{K}$ (or it is asymptotically compact) if it possesses the following property: for every $B \in \mathcal{B}$ such that $\gamma^+(B) \in \mathcal{B}$, each sequence of the form $\{V_{t_k}(x_k)\}_{k=1}^\infty$, where $x_k \in B$ and $t_k \nearrow +\infty$, is precompact.

Here we restrict ourselves to the case of continuous semigroups of class $A\mathcal{K}$. We begin with two elementary propositions concerning continuous semigroups.

### Proposition 3.1
*For every compact set $K$ and $t \in \mathbf{R}^+$ the set $\gamma_{[0,t]}^+(K)$ is compact.*

The proof is evident.

### Proposition 3.2
*If $K$ is compact and $\gamma^+(K)$ is precompact then $\omega(K)$ is a non-empty compact invariant set attracting $K$.*

*Proof* In fact, this proposition was proved in Chapter 2. Denoting $K_1 := [\gamma^+(K)]_X$, we know this is compact and $V_t(K_1) \subset K_1$. So we have a semigroup $\{V_t, t \in \mathbf{R}^+, K_1\}$ of continuous operators $V_t$ acting on a metric space $K_1$. Hence, this semigroup is of class $\mathcal{K}$ and we may apply Theorem 2.1 to obtain the desired result.

∎

Now we pass to the semigroups of class $A\mathcal{K}$.

**Proposition 3.3**

Let $\{V_t, \ t \in \mathbf{R}^+, \ X\}$ be a continuous semigroup of class $AK$. Suppose that $K$ is a compact set such that the $\gamma^+(K)$ is bounded. Then $\gamma^+(K)$ is precompact and thus the statement of Proposition 3.2 is true.

*Proof* Let $y_n$, $n = 1, 2, \ldots$, be an arbitrary sequence of points from $\gamma^+(K)$, i.e. $y_n = V_{t_n}(x_n)$ for some $x_n \in K$ and $t_n \in \mathbf{R}^+$. If the set $\{t_n\}_{n=1}^\infty$ is bounded, the set $\{y_n\}_{n=1}^\infty$ is precompact by Proposition 3.1. If the set $\{t_n\}_{n=1}^\infty$ is unbounded, then we may choose a subsequence $t_{n_j} \nearrow +\infty$ and the set $\{V_{t_{n_j}}(x_{n_j})\}_{j=1}^\infty$ will be precompact due to the fact that the semigroup $\{V_t\}$ is of class $AK$.  ∎

**Proposition 3.4**

Let $\{V_t\}$ be a continuous semigroup of class $AK$. Suppose that $\gamma^+(B) \in \mathcal{B}$ for some $B \in \mathcal{B}$. Then $\omega(B)$ is a non-empty invariant compact set attracting $B$ and $\omega(B)$ is connected if $B$ is connected.

*Proof* As $\gamma^+(x) \in \mathcal{B}$ for each $x \in B$, the $\omega$-limit sets $\omega(x)$, $x \in B$, are non-empty, and hence $\omega(B) \neq \emptyset$. It is evident that $\omega(B)$ is closed and bounded. To prove that it is invariant we have only to check the embedding $\omega(B) \subset V_t(\omega(B))$ since the inverse embedding is always valid provided operators $V_t$ are continuous (see the proof of Theorem 2.1). Choose an arbitrary $y \in \omega(B)$. We know that $y = \lim_{n \to \infty} V_{t_n}(x_n)$ for some $x_n \in B$ and some $t_n \nearrow +\infty$. Obviously, $V_{t_n}(x_n) = V_t(V_{t_n - t}(x_n))$ if $t_n \geq t$. The set $\{V_{t_n - t}(x_n)\}_{t_n \geq t}$ is precompact since $\{V_t\}$ is of class $AK$.

Choose a converging subsequence $\{V_{t_{n_j} - t}(x_{n_j})\}_{j=1}^\infty$ and let $z = \lim_{j \to \infty} V_{t_{n_j} - t}(x_{n_j})$. Clearly $z \in \omega(B)$ and $V_t(z) = y$. Thus the embedding $\omega(B) \subset V_t(\omega(B))$ is established. Hence $\omega(B)$ is invariant. From this it follows that each sequence $\{x_k\}_{k=1}^\infty$ with $x_k \in \omega(B)$ may be represented as $\{x_k = V_k(\overline{x}_k)\}_{k=1}^\infty$ with $\overline{x}_k \in \omega(B)$ and therefore it is precompact, so that $\omega(B)$ is compact.

It remains to prove that $\omega(B)$ attracts $B$. Suppose that it is not true. Then we can choose a sequence $\{V_{t_k}(x_k)\}_{k=1}^\infty$ with $x_k \in B$ and $t_k \nearrow +\infty$ so that $\mathrm{dist}\{\{V_{t_k}(x_k)\}_{k=1}^\infty; \ \omega(B)\} \geq \epsilon > 0$ for some $\epsilon$. The asymptotical compactness of our semigroup $\{V_t\}$ implies the precompactness of the set $\{V_{t_k}(x_k)\}_{k=1}^\infty$. Since all the limit points of the set $\{V_{t_k}(x_k)\}_{k=1}^\infty$ must lie in $\omega(B)$, the distance between $\omega(B)$ and $\{V_{t_k}(x_k)\}_{k=1}^\infty$ is zero. This is a contradiction.

If $B$ is connected then $\omega(B)$ is connected by using the same arguments as in the proof of Theorem 2.1 (v).

∎

### Theorem 3.1

*Let $\{V_t, \ t \in \mathbf{R}^+, \ X\}$ be a continuous bounded and pointwise dissipative semigroup of class $A\mathcal{K}$. Then there exists a non-empty minimal global B-attractor $\mathcal{M}$. $\mathcal{M}$ is compact and invariant. If $X$ is connected then $\mathcal{M}$ is also connected.*

*Proof* By Proposition 2.1 there is a bounded set $B_0$ such that for every compact $K$

$$V_t(\mathcal{O}_{\epsilon(K)}(K)) \subset B_0 \quad \text{for all} \quad t \geq T(K), \tag{3.1}$$

with some $\epsilon(K) > 0$ and $T(K) < +\infty$, and, in addition, $V_t(B_0) \subset B_0$ for all $t \in \mathbf{R}^+$.

Since the semigroup $\{V_t\}$ is bounded (i.e., $\gamma^+(B) \in \mathcal{B}$ for any $B \in \mathcal{B}$), Proposition 3.4 yields that for every $B \in \mathcal{B}$ the $\omega$-limit set $\omega(B)$ is non-empty, compact, invariant and attracts $B$. In particular, so is $\omega(B_0)$. We claim that $\omega(B_0) \equiv \mathcal{M}$. To prove this statement we need only to show that $\omega(B_0)$ attracts each bounded set. But if $B \in \mathcal{B}$ then $B$ is attracted to the compact set $\omega(B) \equiv K$. Hence, $V_t(B) \subset \mathcal{O}_{\epsilon(K)}(K)$ for all $t \geq t_1(B)$. Because of (3.1), $V_{t+t_1(B)}(B) \subset B_0$ for all $t \geq T(K)$. But we know that $B_0$ is attracted to $\omega(B_0)$. Hence, $B$ is attracted to $\omega(B_0)$ as well.

If $X$ is connected then we may choose a bounded connected set $B_1 \supset B_0$. Its $\omega$-limit set $\omega(B_1)$ is connected and it is easy to verify that $\omega(B_1) \equiv \omega(B_0)$.

∎

### Theorem 3.2

*If $\{V_t, \ t \in \mathbf{R}^+, \ X\}$ is a continuous bounded semigroup of class $A\mathcal{K}$ and it has a "good" Lyapunov function $\mathcal{L}: X \to \mathbf{R}$, then all statements of Theorem 2.3 are true for it.*

The proof of the theorem is the same as for Theorem 2.3 if we bear in mind the results of Theorem 3.1.

The following theorem is useful for applications:

### Theorem 3.3

*Suppose that the semigroup $\{V_t, t \in \mathbf{R}^+, M\}$ is defined on a subset $M$ of a Banach space $X$ with a norm $\|\cdot\|_X$. Suppose also that $V_t$ can be decomposed in the sum $W_t + U_t$, where $\{W_t, t \in \mathbf{R}^+, M\}$ is a family of operators such that for any bounded set $B \subset M$*

$$\|W_t(B)\|_X \leq m_1(t) m_2(\|B\|_X), \tag{3.2}$$

*where $m_k : \mathbf{R}^+ \to \mathbf{R}^+$ are continuous and $m_1(t) \to 0$ when $t \to +\infty$, $\|B\|_X := \sup_{x \in B} \|x\|_X$. The $U_t$ are such that the set $U_t(B)$ is precompact for each bounded set $B \subset M$. Then $\{V_t, t \in \mathbf{R}^+, M\}$ belongs to the class AK.*

*Proof* Let $\gamma^+(B) \in \mathcal{B}$. We show that each set $B_1 := \{V_{t_k}(x_k)\}_{k=1}^\infty$, $t_k \nearrow \infty$, $x_k \in B$, can be covered by a finite $\epsilon$-network where $\epsilon$ is any positive number. Let us choose $\ell$ so large that $m_1(\ell) \leq \epsilon [2m_2(\|B\|_X)]^{-1}$ and decompose $B_1$ in the sum $B_1' \cup B_1''$, where $B_1' = \{V_{t_k}(x_k)\}_{k=1}^{k_1}$, $t_k < \ell$, and $B_1'' = \{V_{t_k}(x_k)\}_{k=k_1+1}^\infty$, $t_{k_1+1} \geq \ell$. $B_1''$ is a subset of the set $V_\ell(\gamma^+(B))$ and any element of $V_\ell(\gamma^+(B))$ has the form $W_\ell(x) + U_\ell(x)$, where $x$ is an element of $\gamma^+(B)$. The set $U_\ell(\gamma^+(B))$ may be covered by a finite $\epsilon/2$-network since it is precompact and the norms of the elements of $W_\ell(\gamma^+(B))$ are not larger than $\epsilon/2$. Therefore the set $V_\ell(\gamma^+(B))$ may be covered by a finite $\epsilon$-network. Hence $B_1$ may be covered by a finite $\epsilon$-network as well. ∎

# Afterword

---

Class $\mathcal{K}$ had appeared in connection with the study of the set of all limit-states for the Navier–Stokes equations ([1], 1972). To this class belong the families of solution operators for many problems of parabolic type. Class $A\mathcal{K}$ had arisen for PDE later (in 80's) during the study of some problems of hyperbolic and mixed types. Class $\mathcal{K}$ is a part of class $A\mathcal{K}$, but we have devoted to it the separate Chapter 2 for historical and methodological reasons. Besides these arguments, the results of Chapter 3 about semigroups of class $A\mathcal{K}$ do not cover the results of Chapter 2, since in Chapter 3 we consider (in contrast to Chapter 2) only continuous semigroups. This restriction is not very important for the theory of attractors and the principal facts of the theory are true for semigroups $\{V_t,\, t \in \mathcal{T}^+,\, X\}$ of class $A\mathcal{K}$ with any additive semigroup $\mathcal{T}^+ \subset \mathbf{R}^+$.

Let us formulate, for example, the theorem which generalizes Theorems 2.2 and 3.1.

### Definition

*The semigroup $\{V_t,\, t \in \mathcal{T}^+,\, X\}$ belongs to the class $A\mathcal{K}$ iff for every $B \in \mathcal{B}$ such that $\gamma_{T(B)}^+ \in \mathcal{B}$ for a $T(B) \in \mathcal{T}^+$, each sequence of the form $\{V_{t_k}(x_k)\}_{k=1}^{\infty}$ where $x_k \in B$, $t_k \nearrow \infty$, is precompact.*

### Theorem 3.4

*Let $\{V_t,\, t \in \mathcal{T}^+,\, X\}$ be a point-wise dissipative semigroup of class $A\mathcal{K}$ and suppose that for each $B \in \mathcal{B}$ there exists a $T(B) \in \mathcal{T}^+$*

such that $\gamma^+_{T(B)} \in \mathcal{B}$. *Then there exists a non-empty minimal global B-attractor* $\mathcal{M}$. *It is compact and invariant. If X is connected then* $\mathcal{M}$ *is also connected.*

It is easy to see that if a semigroup has a compact global $B$-attractor then it has all properties indicated in the conditions of Theorem 3.4.

The proofs of this and other theorems extending the theorems of Chapters 2 and 3 to the semigroups $\{V_t,\ t \in \mathcal{T}^+,\ X\}$ of class $A\mathcal{K}$ are close to proofs given here. They will be published elsewhere.

# 4

---

# On dimensions of compact
invariant sets

In this chapter we shall estimate $\dim_H(\mathcal{A})$ and $\dim_f(\mathcal{A})$, i.e. Hausdorff and fractal dimensions of compact invariant sets $\mathcal{A}$, and, as a consequence, of attractors $\mathcal{M}$ provided $X$ is a separable Hilbert space.

Let $K$ be a compact set lying in $X$ and $B_r(x)$ the closed ball of radius $r$ centered at $x$. We associate with every finite covering $\mathcal{U} = \{B_{r_i}(x_i)\}$ of $K$ (i.e. $K \subset \bigcup_i B_{r_i}(x_i)$) the numbers:

$$r(\mathcal{U}) := \max_i r_i \,,$$

$$n(\mathcal{U}) = \text{number of elements in } \mathcal{U} \,,$$

$$m_{\beta,r(\mathcal{U})} := \sum_{i=1}^{n(\mathcal{U})} r_i^\beta$$

and

$$\nu_\beta(\mathcal{U}) := r(\mathcal{U})^\beta n(\mathcal{U}) \,,$$

where $\beta$ is a positive number.

We shall use the following known lemma.

### Lemma 4.1

*Suppose that for the compact set $K$ there is a sequence $\mathcal{U}_s$, $s = 0, 1, \ldots$ of coverings as described above with $r(\mathcal{U}_s) \to 0$, $m_{\beta,r(\mathcal{U}_s)}(\mathcal{U}_s) \to 0$ when $s \to \infty$. Then*

$$\dim_H(K) \leq \beta \,.$$

*If for such a sequence $\nu_\beta(\mathcal{U}_s) \to 0$, then*

$$\dim_f(K) \leq \beta \,.$$

Let $H$ be a separable Hilbert space. We shall use the following notations:

$\mathcal{P}^N$ = an orthogonal projector on an $N$-dimensional subspace $\mathcal{P}^N H$;

$Q^N H$ = an orthogonal complement to $\mathcal{P}^N H$ and $Q^N = I - \mathcal{P}^N$;

$B_r$ = the ball in $H$ of radius $r$ centered at the origin;

$B_r(\mathcal{P}^N)$ and $B_r(Q^N)$ = the analogous balls in the subspaces $\mathcal{P}^N H$ and $Q^N H$, respectively;

$\mathcal{E}(\mathcal{P}^N, \alpha)$ = the ellipsoid in $\mathcal{P}^N H$ centered at the origin with semi-axes $\alpha_1 \geq \alpha_2 \geq \ldots \geq \alpha_N$, where $(\alpha_1, \ldots, \alpha_N) = \alpha$;

$\mathcal{E}(\mathcal{P}^N, \alpha) \oplus B_\delta(Q^N)$ is the set of $v \in H$ such that $v = v_1 + v_2$, where $v_1 \in \mathcal{E}(\mathcal{P}^N, \alpha)$ and $v_2 \in B_\delta(Q^N)$; $v + B = \{v + u \mid u \in B\}$.

In what follows we shall deal with the projectors $\mathcal{P}^N$ and the numbers $\alpha_1, \alpha_2, \ldots$, which depend on the points $v$ of some subsets in $H$. In this case the dependence on $v$ will be denoted by $\mathcal{P}^N(v)$, $\alpha(v)$, etc.

Let $\alpha_1 \geq \alpha_2 \geq \ldots$ be an infinite sequence of nonnegative real numbers. Then $\omega_k(\alpha)$ denotes the product of the first $k$ numbers of the sequence $\alpha$, i.e. $\omega_k(\alpha) = \alpha_1 \ldots \alpha_k$, and $\omega_0(\alpha) := 1$. If the numbers $\alpha_k$ depend on the points $v \in \mathcal{A} \subset H$ we use the notation $\overline{\alpha}_k := \sup_{v \in \mathcal{A}} \alpha_k(v)$ and $\overline{\omega}_k := \sup_{v \in \mathcal{A}} \omega_k(\alpha(v))$. It is obvious that $\alpha_k(v) \leq \omega_k^{1/k}(v)$, $\overline{\alpha}_k \leq \overline{\omega}_k^{1/k}$; $\omega_k^{1/k}(v)$ and $\overline{\omega}_k^{1/k}$ do not increase when $k$ grows.

### Theorem 4.1

*Let $H$ be a Hilbert space, $V$ a continuous mapping from $H$ into $H$, and $\mathcal{A}$ a compact subset of $H$ such that $\mathcal{A} \subset V(\mathcal{A})$. Assume that there are $\alpha_1(v), \ldots \alpha_N(v), \delta(v)$, defined on $\mathcal{A}$ and projectors $\mathcal{P}^N(v)$, $v \in \mathcal{A}$, such that $\alpha_1(v) \geq \ldots \geq \alpha_N(v) \geq \delta(v) > 0$ for all $v \in \mathcal{A}$ and*

$$V((v + B_r) \cap \mathcal{A}) \subset V(v) + r[\mathcal{E}(\mathcal{P}^N(v), \alpha(v)) \oplus B_{\delta(v)}(Q^N(v))] \tag{4.1}$$

*for all $r \leq r_0$ with some $r_0$ and $N \geq 1$.*

If $\overline{\alpha}_1 = \sup_{v \in \mathcal{A}} \alpha_1(v) < +\infty$ and $\overline{\delta} = \sup_{v \in \mathcal{A}} \delta(v) < 1/2$, then

$$\dim_H(\mathcal{A}) \leq \max \left\{ N;\ N\, \frac{\ln[(\sqrt{N}+1)\overline{\omega}_N^{1/N}(\epsilon\overline{\delta})^{-1}]}{\ln\left(1/2\overline{\delta}\sqrt{1+\epsilon^2}\right)} \right\} \qquad (4.2)$$

$$\equiv d_1 ,$$

where $\epsilon$ is an arbitrary number in $(0,1]$ satisfying the inequality $2\overline{\delta}\sqrt{1+\epsilon^2} < 1$

If

$$2\sqrt{2}\left(\sqrt{N}+1\right)\overline{\omega}_N^{1/N} \leq 1 , \qquad (4.3)$$

then $\dim_H(\mathcal{A}) \leq N$.

*Remark* Inequality (4.3) implies $\overline{\delta} \leq \overline{\alpha}_N \leq \overline{\omega}_N^{1/N} < 1/(2\sqrt{2})$ and, hence, in (4.2) we may choose $\epsilon = 1$, and, taking into account (4.3), get $\dim_H(\mathcal{A}) \leq N$.

The next theorem provides an estimate for the fractal dimension of $\mathcal{A}$.

### Theorem 4.2
*Under the assumptions of Theorem 4.1 suppose $\overline{\alpha}_1 < +\infty$ and $\overline{\delta} < 1/(2\sqrt{2})$. Then*

$$\dim_f(\mathcal{A}) \leq N\ln\left[(\sqrt{N}+1)\chi_N\overline{\delta}^{-1}\right] \Big/ \ln\left(\frac{1}{2\sqrt{2}\,\overline{\delta}}\right), \qquad (4.4)$$

*where*

$$\chi_N = \max_{\ell=0,1,\dots,N}\left(\overline{\omega}^{1/N}\overline{\delta}^{-1-\ell/N}\right);\ \overline{\omega}_0 := 1 . \qquad (4.5)$$

*In particular, if*

$$2\sqrt{2}(\sqrt{N}+1)\chi_N \leq 1 , \qquad (4.6)$$

*then $\dim_f(\mathcal{A}) \leq N$.*

*Proof of Theorem 4.1* We construct some "pulverizing coverings" $\mathcal{U}_s$, $s = 0, 1,\dots$ of the set $\mathcal{A}$ with $\rho_s = r(\mathcal{U}_s) \to 0$, $s \to \infty$, and calculate for them the numbers $m_{\beta,\rho_s}(\mathcal{U}_s)$. By choosing $\beta$ sufficiently large we get that $m_{\beta,\rho_s}(\mathcal{U}_s) \to 0$ when $s \to \infty$ and $\dim_H(\mathcal{A}) \leq \beta$.

Let $\mathcal{U}_0$ be a covering of $\mathcal{A}$ by a finite number of balls $v_i + B_{r_i}$ $i = 1,\dots,n(\mathcal{U}_0)$, with $v_i \in \mathcal{A}$ and $r_i \leq r(\mathcal{U}_0) \equiv \rho_0$. The number $m_{\beta,\rho_0}(\mathcal{U}_0) = \sum_{i=1}^{n(\mathcal{U}_0)} r_i^\beta$ corresponds to this covering. The next covering is constructed in the following way: since $\mathcal{A} \subset V(\mathcal{A})$ the collection of sets $V((v_i + B_{r_i}) \cap \mathcal{A})$, $i = 1, \ldots, n(\mathcal{U}_0)$, is a covering of $\mathcal{A}$. Because of the assumption (4.1) we have

$$V((v_i + B_{r_i}) \cap \mathcal{A})$$
$$\subset V(v_i) + r_i \left[ \mathcal{E}(\mathcal{P}^N(v_i), \alpha(v_i)) \oplus B_{\delta(v_i)}(Q^N(v_i)) \right] \quad (4.7)$$
$$\subset V(v_i) + r_i \left[ \pi^N(i) \oplus B_{\delta_i}^{\perp} \right] ,$$

where $\pi^N(i) = \pi(\mathcal{P}^N(v_i), \alpha(v_i))$ is the parallelepiped in $\mathcal{P}^N(v_i)H$ with edges of length $2\alpha_k(i) \equiv 2\alpha_k(v_i)$, $i = 1, \ldots, N$ and $B_{\delta_i}^{\perp}$ is the ball $B_{\delta(v_i)}(Q^N(v_i))$ in $Q^N(v_i)H$. Cover $\pi^N(i)$ by cubes $K^{ij}$, $j = 1, 2, \ldots,$ $n(i)$, with edges of length $2\delta_i \epsilon / \sqrt{N}$, $\delta_i = \delta(v_i)$. The diameter of the set $K^{ij} \oplus B_{\delta_i}^{\perp}$ is equal to $2\delta_i \sqrt{1 + \epsilon^2} \equiv \gamma_i$. Now, there are $v_{ij} \in \mathcal{A}$ such that $[V(v_i) + r_i(K^{ij} \oplus B_{\delta_i}^{\perp})] \cap \mathcal{A} \subset v_{ij} + B_{\gamma_i r_i}$. The collection of the balls $v_{ij} + B_{\gamma_i r_i}$, $i = 1, \ldots, n(\mathcal{U}_0)$, $j = 1, \ldots, n(i)$, is a new covering $\mathcal{U}_1$ of $\mathcal{A}$.

Let us estimate the number $n(i)$. Obviously,

$$n(i) \leq \prod_{k=1}^{N} \left( \left[ \frac{\alpha_k^{(i)}}{\delta_i \epsilon} \sqrt{N} \right] + 1 \right) \leq C_N^N \frac{\omega_N(i)}{(\delta_i \epsilon)^N} , \quad (4.8)$$
$$C_N = \sqrt{N} + 1 .$$

Next, the radii of the balls of the covering $\mathcal{U}_1$ are equal to $\gamma_i r_i = 2\delta_i \sqrt{1 + \epsilon^2} \, r_i \leq \gamma \rho_0$, where $\gamma = 2\bar{\delta}\sqrt{1 + \epsilon^2}$, and $\gamma < 1$ by the choice of $\epsilon$. Hence, we have $\rho_1 \equiv r(\mathcal{U}_1) \leq \gamma \rho_0$. Suppose $\beta \geq N$. Then

$$m_{\beta, \rho_1}(\mathcal{U}_1) = \sum_{i=1}^{n(\mathcal{U}_0)} \sum_{j=1}^{n(i)} (\gamma_i r_i)^{\beta} = \sum_{i=1}^{n(\mathcal{U}_0)} (\gamma_i r_i)^{\beta} n(i)$$
$$\leq C_N^N \overline{\omega}_N \sum_{i=1}^{n(\mathcal{U}_0)} r_i^{\beta} (2\delta_i \sqrt{1 + \epsilon^2})^{\beta} (\delta_i \epsilon)^{-N}$$
$$\leq C_N^N \overline{\omega}_N (2\bar{\delta}\sqrt{1 + \epsilon^2})^{\beta} (\bar{\delta}\epsilon)^{-N} \sum_{i=1}^{n(\mathcal{U}_0)} r_i^{\beta} .$$

Hence, for $\beta \geq N$

$$m_{\beta, \rho_1}(\mathcal{U}_1) \leq C_N^N \overline{\omega}_N (2\bar{\delta}\sqrt{1 + \epsilon^2})^{\beta} (\epsilon\bar{\delta})^{-N} m_{\beta, \rho_0}(\mathcal{U}_0) . \quad (4.9)$$

Now, choose $\beta$ such that

$$\eta = C_N^N \overline{\omega}_N (2\bar{\delta}\sqrt{1 + \epsilon^2})^{\beta} (\epsilon\bar{\delta})^{-N} < 1$$

or, equivalently,

$$\beta > N \ln(C_N \overline{\omega}_N^{1/N} (\epsilon\bar{\delta})^{-1}) \left/ \ln\left( \frac{1}{2\bar{\delta}\sqrt{1 + \epsilon^2}} \right) \right. . \quad (4.10)$$

Then, $m_{\beta,\rho_1}(\mathcal{U}_1) \leq \eta m_{\beta,\rho_0}(\mathcal{U}_0)$ with $\eta < 1$, and $\rho_1 \leq \gamma\rho_0$ with $\gamma < 1$.

Now we start with the covering $\mathcal{U}_1$ and repeat the same procedure to obtain the covering $\mathcal{U}_2$. From $\mathcal{U}_2$ we pass to $\mathcal{U}_3$ and so on. At each step we get $m_{\beta,\rho_{k+1}}(\mathcal{U}_{k+1}) \leq \eta m_{\beta,\rho_k}(\mathcal{U}_k)$ and $\rho_{k+1} \leq \gamma\rho_k$ with the same $\eta$ and $\gamma$ as above. Hence,

$$m_{\beta,\rho_k}(\mathcal{U}_k) \leq \eta^k m_{\beta,\rho_0}(\mathcal{U}_0), \ \rho_k \leq \gamma^k \rho_0, \ \eta < 1, \gamma < 1,$$

for all $k = 1, 2, \ldots$. Therefore, by Lemma 4.1, $\dim_H(\mathcal{A}) \leq \beta$. ∎

*Proof of Theorem 4.2* To prove that $\dim_f(\mathcal{A}) \leq \beta$ it is sufficient to present a sequence of coverings $\mathcal{U}_s$ of the set $\mathcal{A}$ by balls of radius $r_s = r(\mathcal{U}_s)$ such that $r_s \to 0$ and $\nu_\beta(\mathcal{U}_s) := n(\mathcal{U}_s)r_s^\beta \to 0$ when $s \to +\infty$.

Let $\mathcal{U}_0$ be a finite covering of $\mathcal{A}$ by the balls $(v_i + B_{r_0})$, $i = 1, \ldots, n(\mathcal{U}_0)$, with $v_i \in \mathcal{A}$. To obtain the covering $\mathcal{U}_1$ we follow the procedure described in the proof of Theorem 4.1, but with certain modifications: the parallelepipeds $\pi^N(i)$ should be covered by cubes $K^{ij}$, $j = 1, \ldots, n(i)$, with edges of length $2\bar{\delta}/\sqrt{N}$ (thus the diameter of $K^{ij} \oplus B_{\delta_i}^\perp$ is not greater than $\gamma = 2\sqrt{2}\,\bar{\delta}$) The sets $[V(v_i) + r_0(K^{ij} \oplus B_{\delta_i}^{i\perp})] \cap \mathcal{A}$ are embedded into the balls $(v_{ij} + B_{\gamma r_0})$. These balls form the covering $\mathcal{U}_1$.

The number $n(i)$, i.e. the number of cubes $K^{ij}$ needed to cover $\pi^N(i)$, is estimated as follows:

$$n(i) \leq \prod_{k=1}^{N} \left( \left[ \frac{\alpha_k(i)}{\bar{\delta}} \sqrt{N} \right] + 1 \right)$$
$$\leq C_N^N \omega_m(i) / \bar{\delta}^{m(i)} \tag{4.11}$$
$$\leq C_N^N \overline{\omega}_{m(i)} / \bar{\delta}^{m(i)},$$

where

$$m(i) = \begin{cases} \max\{k : 1 \leq k \leq N \text{ and } \alpha_k(i) > \bar{\delta}\}, & \text{if } \bar{\delta} < \alpha_1(i), \\ 0 \quad (\text{in this case } \overline{\omega}_0 := 1), & \text{if } \bar{\delta} \geq \alpha_1(i). \end{cases}$$

It is easy to see that for arbitrary $\beta$

$$n(\mathcal{U}_1)(r(\mathcal{U}_1))^\beta \leq (r_0\gamma)^\beta \sum_{i=1}^{n(\mathcal{U}_0)} n(i)$$

$$\leq r_0^\beta C_N^N \sum_{i=1}^{n(\mathcal{U}_0)} (2\sqrt{2}\,\overline{\delta})^\beta \overline{\omega}_{m(i)}/\overline{\delta}^{m(i)} \qquad (4.12)$$

$$\leq C_N^N (2\sqrt{2})^\beta \overline{\delta}^{\beta-N} \chi_N^N n(\mathcal{U}_0) r_0^\beta ,$$

where $\chi_N$ is defined by (4.5). Now, choose $\beta$ such that

$$\eta = C_N^N (2\sqrt{2})^\beta \overline{\delta}^{\beta-N} \chi_N^N < 1 ,$$

or equivalently,

$$\beta > N \ln(C_N \chi_N \overline{\delta}^{-1}) \Big/ \ln\left(\frac{1}{2\sqrt{2}\,\overline{\delta}}\right) . \qquad (4.13)$$

Then we get from (4.12): $\nu_\beta(\mathcal{U}_1) \leq \eta \nu_\beta(\mathcal{U}_0)$. From the covering $\mathcal{U}_1$ we pass to $\mathcal{U}_2$ and then to $\mathcal{U}_3$ and so on. Obviously, we shall have

$$\nu_\beta(\mathcal{U}_k) \leq \eta^k \nu_\beta(\mathcal{U}_0), \quad r(\mathcal{U}_k) = \gamma^k r(\mathcal{U}_0) = \gamma^k r_0 .$$

Since $\gamma < 1$ and $\eta < 1$, we get

$$\nu_\beta(\mathcal{U}_k) \equiv n(\mathcal{U}_k)(r(\mathcal{U}_k))^\beta \to 0, \quad r(\mathcal{U}_k) = \gamma^k r_0 \to 0$$

when $k \to +\infty$, and therefore (4.4) is proved.

∎

Now we consider the question of verifing the conditions of Theorems 4.1 or 4.2 for some nonlinear differentiable operators $V: H \to H$.

We begin by reminding the reader of some known results concerning linear bounded operators in a Hilbert space.

Let $U$ be a linear bounded operator in $H$. We associate to $U$ the real numbers

$$\alpha_k(U) := \sup_{\substack{\mathcal{L} \subset H \\ \dim \mathcal{L}=k}} \inf_{\substack{\|x\|=1 \\ x \in \mathcal{L}}} \|Ux\| , \qquad (4.14)$$

where sup is taken over all linear $k$-dimensional subspaces $\mathcal{L} \subset H$. The numbers $\alpha_k(U)$ are non-negative defined for all $k = 1, 2, \ldots$, and $\alpha_k(U) \geq \alpha_{k+1}(U)$. Denote $\alpha_\infty(U) := \lim_{k \to \infty} \alpha_k(U)$ and let $\mathcal{T}$ be the set of all indices $i$ such that $\alpha_i(U) > \alpha_\infty(U)$. The set $\mathcal{T}$ may be empty of finite or infinite. Let $B_1$ be the unit ball in $H$ centered about zero. Then

$$U(B_1) \subset \mathcal{E}(\mathcal{P}^{\mathcal{T}}(U); \alpha(U)) \oplus B_{\alpha_\infty(U)}(Q^{\mathcal{T}}(U)) , \qquad (4.15)$$

where $\mathcal{P}^{\mathcal{T}}(U)$ is the orthogonal projector on a certain subspace

$\mathcal{P}^{\mathcal{T}}(U)H \subset H$ whose dimension is equal to the number of elements in $\mathcal{T}$, $Q^{\mathcal{T}}(U) = I \ominus \mathcal{P}^{\mathcal{T}}(U)$, $\mathcal{E}(\mathcal{P}^{\mathcal{T}}(U); \alpha(U))$ is an ellipsoid in $\mathcal{P}^{\mathcal{T}}(U)H$ with the semi-axes $\alpha_i(U)$, $i \in \mathcal{T}$, and $B_{\alpha_\infty(U)}(Q^{\mathcal{T}}(U))$ is the ball of radius $\alpha_\infty(U)$ in $Q^{\mathcal{T}}(U)H$. From (4.15) we deduce that for any $N < \infty$

$$U(B_1) \subset \mathcal{E}(\mathcal{P}^N(U); \alpha(U))$$
$$\oplus B_{\sqrt{\alpha_{N+1}^2(U)+\alpha_\infty^2(U)}}(Q^N(U)) \tag{4.16}$$

where $\mathcal{P}^N(U)$ is the orthogonal projector on a subspace $\mathcal{P}^N(U)H$ of $H$, $\mathcal{E}(\mathcal{P}^N(U); (U))$ is an ellipsoid of $\mathcal{P}^N(U)H$ with semi-axes $\alpha_1(U)$, $\dots$, $\alpha_N(U)$, and $Q^N(U) = I \ominus \mathcal{P}^N(U)$. Instead of (4.16) we shall use

$$U(B_1) \subset \mathcal{E}(\mathcal{P}^N(U); \sqrt{2}\alpha(U)) \oplus B_{\sqrt{2}\alpha_N(U)}(Q^N(U)) \tag{4.17}$$

We need the following lemma (see [3]):

**Lemma 4.2**

Let $\alpha_1, \dots, \alpha_N$, $M$, $k_0$ be real numbers satisfying the following conditions: $0 \leq \alpha_N \leq \dots \leq \alpha_1 \leq M$ and $\omega_N(\alpha) = \alpha_1 \dots \alpha_N \leq k_0$, where $0 < k_0 \leq M^N$. Let $m := k_0 M^{1-N}$. Define $\alpha_1', \dots, \alpha_N'$ according to the rule:

$$\alpha_i' := \alpha_i \quad \text{if} \quad \alpha_i \geq m, \quad \text{and} \quad \alpha_i' := m \quad \text{if} \quad \alpha_i < m.$$

Then $\alpha_i' \geq \alpha_i$, $i = 1, \dots, N$; $\alpha_N' \geq m$, $\alpha_1' \leq M$, and $\omega_N(\alpha') \leq k_0$. If, in addition, $\omega_0(\alpha) := 1$ and for all $\ell = 0, 1, \dots, N$ we have

$$j_\ell(\alpha) := \omega_\ell(\alpha)k_0^{1-\ell/N} \leq \tilde{k}$$

with some $\tilde{k}$, then $j_\ell(\alpha') \leq \tilde{k}$ for $\ell = 0, 1, \dots, N$.

*Proof* If $\alpha_1 \geq m$ then $\alpha_1' = \alpha_1$, hence $\alpha_1' \leq M$. If $\alpha_1 < m$ then $\alpha_1' = m$. But $m \leq M$ and hence again $\alpha_1' \leq M$. In any case, $\omega_N(\alpha') = \omega_q(\alpha)m^{N-q}$ for some integer $q$, $0 \leq q \leq N$ ($\omega_0(\alpha) = 1$ by definition). Indeed, if $\alpha_1 \geq \dots \geq \alpha_q \geq m > \alpha_{q+1} \geq \dots \geq \alpha_N$ then $\alpha_1' = \alpha_1, \dots, \alpha_q' = \alpha_q$ and $\alpha_{q+1}' = \dots = \alpha_N' = m$; if $\alpha_1 \geq \dots \geq \alpha_N \geq m$ then all $\alpha_i' = \alpha_i$, $j_\ell(\alpha') = j_\ell(\alpha) \leq \tilde{k}$ and $q = N$; if $m > \alpha_1$ then all $\alpha_i' = m$ and $q = 0$.

For $q \leq N - 1$ we have $\omega_N(\alpha') \leq M^q m^{N-q} \leq k_0$. Now consider $j_\ell(\alpha') = \omega_\ell(\alpha')k_0^{1-\ell/N}$ for $q \leq N - 1$. If $\alpha_\ell \geq m$ then $\alpha_1' = \alpha_1, \dots, \alpha_\ell' = \alpha_\ell$ and $\omega_\ell(\alpha') = \omega_\ell(\alpha)$. Hence $j_\ell(\alpha') = j_\ell(\alpha) \leq \tilde{k}$. If $\alpha_\ell < m$ then $\alpha_1 \geq \dots \geq \alpha_q \geq m > \alpha_{q+1} \geq \dots \geq \alpha_\ell$ or $m > \alpha_1$ (and $q = 0$).

In these cases $\omega_\ell(\alpha') = \omega_q(\alpha)m^{\ell-q}$ and therefore

$$j_\ell(\alpha') = \omega_q(\alpha)m^{\ell-q}k_0^{1-\ell/N} = j_q(\alpha)(k_0/M^N)^{(\ell-q)(1-1/N)}$$

$$\leq j_q(\alpha) \leq \tilde{k} .$$

∎

Now we return to the embedding (4.17). Because of Lemma 4.2 we have

$$U(B_1) \subset \mathcal{E}(\mathcal{P}^N(U); \sqrt{2}\alpha'(U)) \oplus B_{\sqrt{2}\alpha'_N(U)}(Q^N(U)) , \quad (4.18)$$

with the numbers $\alpha'_1(U), \ldots, \alpha'_N(U)$ computed from $\alpha_1(U), \ldots, \alpha_N(U)$, $M$ and $k_0$ as was described in Lemma 4.2. Here the only restrictions on $M$ and $k_0$ are: $M \geq \alpha_1(U)$, $\omega_N(\alpha(U)) \leq k_0$ and $0 < k_0 \leq M^N$.

Let $\mathcal{A}$ be a compact set in $H$ and $V: H \to H$ a continuous operator. Suppose that $V$ is *uniformly differentiable on* $\mathcal{A}$, i.e. for every $v \in \mathcal{A}$ there exists a linear bounded operator $U(v)$ such that for all $v_1 \in B_{r_1}(v)$,

$$\|V(v_1) - V(v) - U(v)(v_1 - v)\| \leq \gamma(\|v_1 - v\|)\|v_1 - v\| , (4.19)$$

where $\sup_{v \in \mathcal{A}} \|U(v)\| \leq M < +\infty$ and the function $\gamma(\tau)$ is continuous on the interval $[0, r_1]$ and $\gamma(0) = 0$. The number $r_1$ and $\gamma(\tau)$ do not depend on $v \in \mathcal{A}$.

For this linear bounded operator $U(v)$ the numbers $\alpha_k(U(v))$ are computed according to (4.14). We choose $M$ and $k_0$ so that

$$M \geq \sup_{v \in \mathcal{A}} \alpha_1(U(v)) , \quad k_0 \geq \sup_{v \in \mathcal{A}} \omega_N(\alpha(U(v)))$$

and $0 < k_0 \leq M^N$.

Since $\inf_{v \in \mathcal{A}} \alpha'_N(U(v)) \geq m = k_0 M^{1-N}$ there is $r_0 > 0$ such that the $r_0$-neighbourhood of the set

$$\mathcal{E}(\mathcal{P}^N(U(v)); \sqrt{2}\alpha'(U(v))) \oplus B_{\sqrt{2}\alpha'_N(U(v))}(Q^N(U(v)))$$

lies entirely in the similar set but with the parameters $2\alpha'_k(U(v))$ instead of $\sqrt{2}\alpha'_k(U(v))$. Thus combining (4.18) and (4.19) we obtain

**Lemma 4.3**

*Suppose that $V$ is uniformly differentiable on $\mathcal{A}$. Then there is $r_0 > 0$ such that for all $r \leq r_0$ and all $v \in \mathcal{A}$*

$$V((v + B_r) \cap \mathcal{A})$$

$$\subset V(v) + r\big[\mathcal{E}(\mathcal{P}^N(U(v)); 2\alpha'(U(v))) \quad (4.20)$$

$$\oplus B_{2\alpha'_N(U(v))}(Q^N(U(v)))\big] .$$

Theorem 4.1 and Lemmas 4.2 and 4.3 yield

### Theorem 4.3

*Let $V$ be uniformly differentiable on $\mathcal{A}$, $\mathcal{A} \subset V(\mathcal{A})$ and for differentials $U(v)$, $v \in \mathcal{A}$, of $V$ the inequality*

$$4\sqrt{2}(\sqrt{N}+1)\overline{k}_0^{1/N} \leq 1, \quad N \geq 1,$$

*where $\overline{k}_0 := \sup_{v \in \mathcal{A}} \omega_N(\alpha(U(v))$, holds. Then $\dim_H(\mathcal{A}) \leq N$.*

*Proof* Let $k_0$ be a positive number in the interval

$$[\overline{k}_0, \{4\sqrt{2}(\sqrt{N}+1)\}^{-N}],$$

$M := \max\{k_0^{1/N}; \sup_{v \in \mathcal{A}} \alpha_1(U(v))\}$ and $\alpha_i'(U(v))$, $i = 1, \ldots, N$, be numbers constructed with the help of $\alpha_i(U(v))$, $k_0$ and $M$ as described in Lemma 4.2. Then the embeddings (4.20) are true and we may use the second statement of Theorem 4.1. In fact, (4.1) holds with $\alpha_i(v) := 2\alpha_i'(U(v))$, $i = 1, \ldots, N$; $\delta(v) := 2\alpha_N'(U(v))$ and $\alpha_1(v) \geq \alpha_2(v) \geq \ldots \geq \alpha_N(v) = \delta(v) > 0$. Moreover, (4.3) is also fulfilled since $\overline{\omega}_N^{1/N} := \sup_{v \in \mathcal{A}} \omega_N^{1/N}(\alpha(v)) \leq 2\sup_{v \in \mathcal{A}} \omega_N^{1/N}(\alpha'(U(v)) \leq 2k_0^{1/N} \leq [2\sqrt{2}(\sqrt{N}+1)]^{-1}$. So $\dim_H(\mathcal{A}) \leq N$. ∎

### Theorem 4.4

*Let $V$ have the same properties as in Theorem 4.3 and suppose that*

$$4\sqrt{2}(\sqrt{N}+1) \max_{\ell=0,\ldots,N} \left\{ \sup_{v \in \mathcal{A}} \omega_\ell(\alpha(U(v)))\overline{k}_0^{1-\ell/N} \right\}^{1/N} \leq 1,$$

*where $N \geq 1$, $\overline{k}_0 := \sup_{v \in \mathcal{A}} \omega_N(\alpha(U(v)))$, hold. Then $\dim_f(\mathcal{A}) \leq N$.*

*Proof* If $\overline{k}_0$ is positive we choose $k_0 = \overline{k}_0$,

$$M = \max\{k_0^{1/N}; \sup_{v \in \mathcal{A}} \alpha_1(U(v))\}$$

and take $\alpha_i'(U(v))$, $i = 1, \ldots, N$ corresponding to these numbers and to the numbers $\alpha_i(U(v))$, $i = 1, \ldots, N$. Then the embeddings (4.20) are true and we may use the second statement of Theorem 4.2. Actually, (4.1) holds with $\alpha_i(v) := 2\alpha_i'(U(v))$, $i = 1, \ldots, N$; $\delta(v) := 2\alpha_N'(U(v))$ and $\alpha_1(v) \geq \ldots \geq \alpha_N(v) = \delta(v) > 0$. Now we have to verify the condition (4.6) with $\overline{\omega}_\ell = \sup_{v \in \mathcal{A}} \omega_\ell(\alpha(v)) = $

$2^\ell \sup_{v \in \mathcal{A}} \omega_\ell(\alpha'(U(v)))$ and $\overline{\delta} = \sup_{v \in \mathcal{A}} \delta(v) = 2 \sup_{v \in \mathcal{A}} \alpha'_N(U(v))$.
According to Lemma 4.2

$$\overline{\delta} \leq 2 \sup_{v \in \mathcal{A}} \omega_N^{1/N}(\alpha'(U(v)) \leq 2k_0^{1/N} \text{ and}$$

$$\chi_N := \max_{\ell = 0, 1, \ldots, N} \overline{\omega}_\ell^{1/N} \overline{\delta}^{1 - \ell/N}$$

$$\leq 2 \sup_{v \in \mathcal{A}} \{ \max_{\ell = 0, \ldots, N} \omega_\ell(\alpha'(U(v))) k_0^{1 - \ell/N} \}^{1/N}$$

$$\leq 2 \sup_{v \in \mathcal{A}} \{ \max_{\ell = 0, \ldots, N} \omega_\ell(\alpha(U(v))) k_0^{1 - \ell/N} \}^{1/N} .$$

The inequality $2\sqrt{2}(\sqrt{N} + 1)\chi_N \leq 1$ is therefore guaranteed by the
condition of Theorem 4.4 and so $\dim_f(\mathcal{A}) \leq N$.

If $\overline{k}_0 = 0$ we take $M = \max\{1; \hat{\alpha}_1\}$, where $\hat{\alpha}_1 := \sup_{v \in \mathcal{A}} \alpha_1(U(v))$,
and a number $k_0 \in (0, (\beta M)^{-N^2}]$, where $\beta := 4\sqrt{2}(\sqrt{N} + 1)$, and
construct $\alpha'_i(U(v))$, $i = 1, \ldots, N$, the corresponding $\alpha_i(U(v))$, $i = 1, \ldots, N$, and these $k_0$ and $M$.

Then

$$\beta^{-N} \geq \max_{\ell = 0, \ldots, N-1} \{ M^\ell k_0^{1 - \ell/N} \}$$

$$\geq \max_{\ell = 0, \ldots, N} \{ \sup_{v \in \mathcal{A}} \omega_\ell(\alpha(U(v))) k_0^{1 - \ell/N} \} .$$

According to Lemma 4.2

$$\max_{\ell = 0, \ldots, N} \{ \omega_\ell(\alpha(U(v))) k_0^{1 - \ell/N} \}$$

$$\geq \max_{\ell = 0, \ldots, N} \{ \omega_\ell(\alpha'(U(v))) k_0^{1 - \ell/N} \} ,$$

for all $v \in \mathcal{A}$.

Therefore

$$\chi_N = \max_{\ell = 0, \ldots, N} \{ \overline{\omega}_N^{1/N} \overline{\delta}^{1 - \ell/N} \}$$

$$\leq 2 \max_{\ell = 0, \ldots, N} \{ \sup_{v \in \mathcal{A}} \omega_\ell(\alpha'(U(v))) k_0^{1 - \ell/N} \}^{1/N}$$

$$\leq [2\sqrt{2}(\sqrt{N} + 1)]^{-1} .$$

So the condition (4.6) is fulfilled and $\dim_f(\mathcal{A}) \leq N$. ∎

When $\mathcal{A}$ is invariant, i.e. $V(\mathcal{A}) = \mathcal{A}$, the following results hold.

### Theorem 4.5
*Suppose that $V$ is a uniformly differentiable operator on a
compact invariant set $\mathcal{A}$, and its differentials $U(v)$, $v \in \mathcal{A}$, satisfy the*

*condition:*

$$\sup_{v \in \mathcal{A}} \omega_N(\alpha(U(v))) := \overline{k}_0 < 1 . \tag{4.21}$$

*Then* $\dim_H(\mathcal{A}) \leq N$.

### Theorem 4.6

*Under the assumptions of Theorem 4.5 let*

$$\max_{\ell=0,\dots,N} \{\sup_{v \in \mathcal{A}} \omega_\ell(\alpha(U(v)))\overline{k}_0^{1-\ell/N}\} < 1 , \tag{4.22}$$

*where* $\omega_0(\alpha) := 1$.

*Then* $\dim_f(\mathcal{A}) \leq N$.

*Proof of Theorem 4.5* Theorem 4.5 is deduced from Theorem 4.3 applied to the operator $V_p$, the $p$-th power of the operator $V$ which is denoted now by $V_1$.

It is known that $\omega_k(\alpha(U_1 \cdot U_2)) \leq \omega_k(\alpha(U_1))\omega_k(\alpha(U_2))$ for all $k \geq 1$ and arbitrary bounded linear operators $U_1$, $U_2$. Since the differentials $U_p(v)$, $v \in \mathcal{A}$ of the operators $V_p$ have the form $U(v_1)\dots U(v_p)$, where $v_k \in \mathcal{A}$, and $U(v_k)$ is the differential of $V_1$, the following inequalities hold:

$$\sup_{v \in \mathcal{A}} \omega_k(\alpha(U_p(v))) := \overline{\omega}_k(p) \leq (\overline{\omega}_k(1))^p .$$

By the assumption (4.21) we have $\overline{\omega}_N(p) \leq \overline{k}_0^p$. Choose the integer $p$ large enough that $4\sqrt{2}(\sqrt{N} + 1)\overline{k}_0^{p/N} \leq 1$. Applying Theorem 4.3 (for the operator $V = V_p$) we obtain $\dim_H(\mathcal{A}) \leq N$.

∎

In a similar way Theorem 4.6 is deduced from Theorem 4.4.

Theorems 4.5 and 4.6 may be used to estimate the Hausdorff and fractal dimensions of compact sets $\mathcal{A}$ which are invariant with respect to evolution operators $V_t: H \to H$, $t \in \mathbf{R}^+$, of the problems

$$\partial_t v(t) = \Phi(v(t)) , \quad v|_{t=0} = v_0 , \tag{4.23}$$

where $\Phi$ is a nonlinear (generally unbounded) operator, subjected to some restrictions.

Suppose we know that problem (4.23), for every $v_0 \in H$, has a unique solution defined on $\mathbf{R}^+$, and the solution operators $V_t$, $t \in \mathbf{R}^+$, form a semigroup in $H$. Let $\mathcal{A}$ be a compact set invariant with respect to this semigroup. In what follows we deal exclusively with the semigroup $V_t$ restricted to the set $\mathcal{A}$.

Let $U(t, v_0)$ be the differential of the operator $V_t$ at a point $v_0 \in H$. It is known that $U(t, v_0)$ is the solution operator for the linear problem

$$\partial_t u(t) = L(t)u(t) , \quad u|_{t=0} = u_0 , \qquad (4.24)$$

where $L(t) = L(t, v_0)$ is the differential of $\Phi(.)$ at the point $v(t) = V_t(v_0)$.

Usually the investigation of the nonlinear problem (4.23) begins with the study of some linearizations of problem (4.23) similar to (4.24). As a final result one gets the existence of the operators $V_t$, $t \in \mathbf{R}^+$, and of its differentials. In Part II we shall consider some problems of this kind.

Now we are interested in estimating from above the numbers $\omega_n$ for the operators $U(t, v_0)$. We consider two classes of problem (4.23). For the first class, the operators $V_t$ and $U(t, v_0)$ will be compact for $t > 0$; this very property distinguishes the problems of parabolic type. For the second class, these operators are merely continuous and bounded (for example, in the case of problems of hyperbolic type). In either case, the numbers $\omega_n$ may be estimated as follows.

It is known that for every linear bounded operator $U$, the number $\omega_n(U) \equiv \omega_n(\alpha(U)) = \alpha_1(U) \cdot \ldots \cdot \alpha_n(U)$ (where $\alpha_k(U)$ are defined in (4.14)) is the norm of the operator $\Lambda^n(U)$ in the Hilbert space $\Lambda^n(H)$. The space $\Lambda^n(H)$ consists of the elements $u_1 \wedge \ldots \wedge u_n$ with $u_k \in H$, and the inner product in $\Lambda^n(H)$ is defined by $(u_1 \wedge \ldots \wedge u_n, v_1 \wedge \ldots \wedge v_n) = \det((u_i, v_j))$, where $(\cdot, \cdot)$ is the inner product in $H$. We denote the norm of $u_1 \wedge \ldots \wedge u_n$ by $\|u_1 \wedge \ldots \wedge u_n\|$.

The operator $\Lambda^n(U)$ acts in $\Lambda^n(H)$ according to the rule

$$\Lambda^n(U)(u_1 \wedge \ldots \wedge u_n) = Uu_1 \wedge \ldots \wedge Uu_n .$$

It is known also (see [9], [10], etc.) that

$$\sum_{k=1}^{n} \left( u_1 \wedge \ldots \wedge U u_k \ldots \wedge u_n , u_1 \wedge \ldots \wedge u_k \ldots \wedge u_n \right) \qquad (4.25)$$
$$= \|u_1 \wedge \ldots \wedge u_n\|^2 \operatorname{tr}(\mathcal{P}^n(\vec{u})U\mathcal{P}^n(\vec{u})) ,$$

where $\mathcal{P}^n(\vec{u})$ is the orthogonal projector on the span $\{u_1, \ldots, u_n\}$ in $H$. Since $\omega_n(U)$ is the norm of $\Lambda^n(U)$ in $\Lambda^n(H)$, we have $\omega_n(U_1 U_2) \leq \omega_n(U_1)\omega_n(U_2)$; this inequality was used in the proofs of Theorems 4.5 and 4.6.

Let us take $n$ linearly independent elements $u_i(0) \in H$ and let $u_i(t)$ be the corresponding solutions of (4.24) with $u_i|_{t=0} = u_i(0)$. Obviously, $u_i(t)$ are linearly independent for all $t > 0$.

Let $\mathcal{P}^n(\vec{u}(t)) := \mathcal{P}^n(t)$ be the orthogonal projector on the span of $\{(u_1(t), \ldots, u_n(t)\}$. From (4.24) we get

$$\partial_t u_i(t) = L_{(n)}(t)u_i(t) + \partial_t \mathcal{P}^n(t)u_i(t), \quad i = 1, \ldots, n, \quad (4.26)$$

where $L_{(n)}(t) = \mathcal{P}^n(t)L(t)\mathcal{P}^n(t)$. Set $G_n(t) := \|u_1(t) \wedge \ldots \wedge u_n(t)\|^2$. By (4.26) we have the Liouville formula:

$$\begin{aligned}
\frac{1}{2}\partial_t G_n(t) &= G_n(t)\operatorname{tr}[L_{(n)}(t) + \partial_t \mathcal{P}^n(t)] \\
&= G_n(t)\operatorname{tr} L_{(n)}^c(t),
\end{aligned} \quad (4.27)$$

where $L_{(n)}^c(t) = \frac{1}{2}[L_n(t) + L_n^*(t)]$. Here we have used the identities $\operatorname{tr}(\partial_t \mathcal{P}^n(t)) = \partial_t \operatorname{tr} \mathcal{P}^n(t)$ and $\operatorname{tr} \mathcal{P}^n(t) = n$, valid for any orthogonal projector $\mathcal{P}^n(t)$.

Formula (4.27) is proved by the same procedure as the usual Liouville formula for a system of $n$ ordinary differential equations in $\mathbf{R}^n$. (Although the system (4.26) is not a system of ODE, since $\partial_t u_i(t)$ and $\partial_t \mathcal{P}^n(t)u_j(t)$ might not lie in $\mathcal{P}^n(t)H$.) Integrating (4.27) we obtain

$$\begin{aligned}
&\|u_1(t) \wedge \ldots \wedge u_n(t)\| \\
&\qquad = \|u_1(0) \wedge \ldots \wedge u_n(0)\| \exp\Big\{\int_0^t \operatorname{tr}[L_{(n)}^c(\tau)]d\tau\Big\},
\end{aligned} \quad (4.28)$$

where $u_i(t) = U(t)u_i(0)$ and $U(t)$ is the solution operator for (4.24). Hence the following estimate holds:

$$\begin{aligned}
\omega_n(U(t)) &= \|\Lambda^n(U(t))\| \\
&= \sup_{u_k(0)\in H} \frac{\|U(t)u_1(0) \wedge \ldots \wedge U(t)u_n(0)\|}{\|u_1(0) \wedge \ldots \wedge u_n(0)\|} \\
&\leq \sup_{u_k(0)\in H} \exp\Big\{\int_0^t \operatorname{tr}(\mathcal{P}^n(\tau)L_{(n)}^c(\tau)\mathcal{P}^n(\tau))d\tau\Big\},
\end{aligned} \quad (4.29)$$

where $\mathcal{P}^n(t) = \mathcal{P}^n(\vec{u}(t))$ is the orthogonal projector on the span $\{(U(t)u_1(0), \ldots, U(t)u_n(0)\}$.

Usually when the existence problem for (4.24) is investigated, the "strength" of the operators $L(t)$ is "measured" by some fixed self-adjoint positively defined operator $A$, whose inverse $A^{-1}$ is compact. Here we shall follow the same line.

Let $H_s$, $s \in \mathbf{R}$, be a scale of Hilbert spaces with the inner products $(u, v)_s \equiv (A^{s/2}u, A^{s/2}v)$ and the norms $\|u\|_s = (u, u)_s^{1/2}$. The original $H$ coincides with $H_0$, $(\cdot, \cdot)_0 = (\cdot, \cdot)$ and $\|\cdot\|_0 = \|\cdot\|$.

For some problems of parabolic type the operators $A^{-1/2}L(t)A^{-1/2}$

are bounded and for almost all $t \geq 0$

$$(L^c(t)u, u) \leq -h_1(t)\|u\|_1^2 + \sum_{k=1}^m h_{s_k}(t)\|u\|_{s_k}^2 , \qquad (4.30)$$

for any $u \in H_1$, where $s_k \leq 0$, $h_1, h_{s_k} \in L_{1,loc}(\mathbf{R})$ and $h_{s_k}(t) \geq 0$, $h_1(t) \geq h_1 > 0$. These inequalities make it possible to estimate the right hand side of (4.29).

Explicitly, if $\mathcal{P}^n$ is an orthogonal projector on the span $\{\phi_1, \ldots, \phi_n\}$, where $\phi_i \in H$ and $(\phi_i, \phi_j) = \delta_{ij}$, the following estimates hold:

$$
\begin{aligned}
\mathrm{tr}(\mathcal{P}^n L^c(t)\mathcal{P}^n) &= \sum_{i=1}^n (L^c(t)\phi_i, \phi_i) \\
&\leq -h_1(t) \sum_{i=1}^n (A\phi_i, \phi_i) \\
&\quad + \sum_{k=1}^m h_{s_r}(t) \sum_{i=1}^n (A^{s_k}\phi_i, \phi_i) .
\end{aligned}
\qquad (4.31)
$$

Since $A$ is self-adjoint with positive eigenvalues $\lambda_k(A)$ and $\lambda_k(A) \nearrow +\infty$ as $k \to +\infty$, in view of the min-max principle we have

$$\sum_{i=1}^n (A^s\phi_i, \phi_i) \leq \mathrm{Sp}_n A^s, \quad \text{if } s \leq 0, \quad \text{and}$$

$$\sum_{i=1}^n (A^s\phi_i, \phi_i) \geq \mathrm{Sp}_n A^s, \quad \text{if } s \geq 0 .$$

Therefore, (4.31) implies

$$
\begin{aligned}
\mathrm{tr}(\mathcal{P}^n L^c(t)\mathcal{P}^n) \leq & -h_1(t)\,\mathrm{Sp}_n A \\
& + \sum_{k=1}^m h_{s_k}(t)\,\mathrm{Sp}_n A^{s_k}, s_k \leq 0 .
\end{aligned}
\qquad (4.32)
$$

For the class of problems under consideration (problems of parabolic type), the solutions $u_k(t) = U(t)u_k(0)$, $u_k(0) \in H$, for almost all $t \in \mathbf{R}^+$ lie in $H_1$, $\int_0^t \|u_k(\tau)\|_1^2 d\tau < +\infty$ and $u_k \in C(\mathbf{R}^+, H)$. Hence $\mathcal{P}^n(\vec{u}(t))$ satisfies for almost all $t$ the assumptions used in the proof of (4.32). Therefore, (4.32) and (4.29) yield

$$
\begin{aligned}
\omega_n(U(t)) \leq \exp\{ & -\int_0^t h_1(\tau)d\tau\,\mathrm{Sp}_n A \\
& + \sum_{k=1}^m \int_0^t h_{s_k}(\tau)d\tau\,\mathrm{Sp}_n A^{s_k}\}
\end{aligned}
\qquad (4.33)
$$

for all $t \in \mathbf{R}^+$. The terms with $s_k \neq 0$ appear in some applications.

As for problems of hyperbolic type (and many others) the operators $V_t$ and $U(t, v_0)$ are merely continuous and bounded. Instead of (4.30) in that case, for any $u \in H$, we have:

$$(L^c(t)u, u) \leq -h_0(t)\|u\|^2 + \sum_{k=1}^{m} h_{s_k}(t)\|u\|_{s_k}^2 , \qquad (4.34)$$

where $s_k < 0$; $h_0, h_{s_k} \in L_{1,loc}(\mathbf{R})$, $h_{s_k}(t) \geq 0$, $h_0(t) \geq h_0 > 0$.

As in the proof of (4.33) we obtain

$$\begin{aligned} \omega_n(U(t)) \leq \exp\{ &- n \int_0^t h_0(\tau)\mathrm{d}\tau \\ &+ \sum_{k=1}^{m} \int_0^t h_{s_k}(\tau)\mathrm{d}\tau \, \mathrm{Sp}_n \, A^{s_k}\} . \end{aligned} \qquad (4.35)$$

We now summarize the results:

### Theorem 4.7

*Let the operators $L(t)$ in (4.24) satisfy the inequalities (4.30) or (4.34). Then the corresponding estimates (4.33) or (4.35) (for $t \geq 0$ and $n \geq 1$) hold.*

From Theorems 4.5 and 4.7 follows:

### Theorem 4.8

*Let $\{V_t, t \in \mathbf{R}^+, H\}$, be a semigroup of solution operators for problem (4.23), and $\mathcal{A}$ a compact set invariant with respect to $V_t$. Let $V_t$ and $\Phi(\cdot)$ be uniformly differentiable on $\mathcal{A}$ and let $L(t, v_0)$ be a differential of $\Phi$ at the point $V_t(v_0)$, $v_0 \in \mathcal{A}$. Suppose that $L(t, v_0)$, $v_0 \in \mathcal{A}$, satisfies the inequalities (4.30) or (4.34) with the functions $h_i(t)$ independent of $v_0 \in \mathcal{A}$.*

*Then $\dim_H(\mathcal{A}) \leq N$ where $N$ is the minimal positive integer such that the expression in the braces on the right hand side of (4.33) (respectively, (4.35)) is negative for some $t > 0$.*

In order to majorize $\dim_f(\mathcal{A})$ we have to estimate from above

$$j_n(t) \equiv \overline{\omega}_n(t)\overline{\omega}_N^{1-n/N}(t) , n = 0, 1, \ldots, N ,$$

where $\overline{\omega}_n(t) \equiv \sup_{v_0 \in \mathcal{A}} \omega_n(U(t, v_0))$. Denote by $U(t)$ the operator $U(t, v_0)$ for some $v_0 \in \mathcal{A}$, and $\overline{h}_s(t) := t^{-1} \int_0^t h_s(\tau)\mathrm{d}\tau$. Suppose that

$$\begin{aligned} \mathrm{Sp}_n \, A &\geq c_1 n^{1+\gamma}, \quad \gamma > 0, \quad c_1 > 0, \\ \mathrm{Sp}_n \, A^s &\leq c_s n^{1+s}, \quad s \leq 0, \quad c_0 = 1 . \end{aligned} \qquad (4.36)$$

Such inequalities hold for the Laplace operator, for the Stokes operator (with any of the classical boundary conditions) and for many other operators in mathematical physics.

Take, for example, one of the summands from $\sum_{k=1}^{m}$ in (4.30) (or (4.34)), say $h_s(t)\|u\|_s^2$, $s \leq 0$. From (4.36) it follows that, for each $n \geq 1$,

$$\omega_n(U(t)) \leq \exp t[-\overline{h}_1(t)c_1 n^{1+\gamma} + \overline{h}_s(t)c_s n^{1+s}]$$
$$\equiv \Psi_n(t) . \tag{4.37}$$

This $\Psi_n(t)$ majorizes $\overline{\omega}_n(t)$ as well. Therefore,

$$j_n(t) \equiv \overline{\omega}_n(t)\overline{\omega}_N^{1-n/N}(t) \leq \Psi_n(t)\Psi_N(t)^{1-n/N} ,$$
$$\frac{1}{t}\ln j_n(t) \leq \frac{1}{t}[\ln \Psi_n(t) + (1 - \frac{n}{N})\ln \Psi_N(t)]$$
$$\equiv -\overline{h}_1(t)c_1 n^{1+\gamma} + \overline{h}_s(t)c_s n^{1+s}$$
$$+ (1 - \frac{n}{N})[-\overline{h}_1(t)c_1 N^{1+\gamma} + \overline{h}_s(t)c_s N^{1+s}]$$
$$= -\overline{h}_1(t)c_1(n^{1+\gamma} + N^{1+\gamma} - nN^\gamma)$$
$$+ \overline{h}_s(t)c_s(n^{1+s} + N^{1+s} - nN^s) .$$

By the Young inequality we have
$$n^{1+\gamma} + N^{1+\gamma} - nN^\gamma$$
$$\geq n^{1+\gamma} + N^{1+\gamma} - \frac{1}{1+\gamma}n^{1+\gamma} - \frac{\gamma}{1+\gamma}N^{1+\gamma}$$
$$= \frac{1}{1+\gamma}N^{1+\gamma} + \frac{\gamma}{1+\gamma}n^{1+\gamma} \geq \frac{1}{1+\gamma}N^{1+\gamma} ,$$

for $\gamma > 0$; equality also holds in the case $\gamma = 0$.

The second sum is estimated from above for $n \in [1, N]$ (remember that $s \leq 0$):
$$n^{1+s} + N^{1+s} - nN^s \leq N^{1+s} .$$

Thus we obtain
$$\frac{1}{t}\ln j_n(t) \leq -\overline{h}_1(t)c_1\frac{1}{1+\gamma}N^{1+\gamma} + \overline{h}_s(t)c_s N^{1+s} ,$$
$$1 \leq n \leq N - 1 , \tag{4.38}$$

and besides,
$$\frac{1}{t}\ln j_N(t) = \frac{1}{t}\ln j_0(t) = \frac{1}{t}\ln \Psi_N(t)$$
$$\equiv -\overline{h}_1(t)c_1 N^{1+\gamma} + \overline{h}_s(t)c_s N^{1+s} . \tag{4.39}$$

In view of Theorem 4.8, $\dim_H(\mathcal{A}) \leq N$, where $N$ has been chosen in such a way that

$$-\overline{h}_1(t)c_1 N^{\gamma-s} + \overline{h}_s(t)c_s < 0 \qquad (4.40)$$

for some $t > 0$.

In order to get $\dim_f(\mathcal{A}) \le N$ (see Theorem 4.6) it is sufficient to choose $t$ and $N$ so that

$$-\overline{h}_1(t)c_1 \frac{1}{1+\gamma} N^{\gamma-s} + \overline{h}_s(t)c_s < 0 \qquad (4.41)$$

Remember that $\gamma > 0$ and $s \le 0$ in (4.40) and (4.41) (see (4.36)).

The above calculations also show that in the case (4.34) $\dim_H(\mathcal{A}) \le \dim_f(\mathcal{A}) \le N$, where $N$ is chosen in such a way that

$$-\overline{h}_0(t)N^{-1} + \overline{h}_s(t)c_s < 0 , \quad s < 0$$

for some $t > 0$.

When there are several summands in $\sum_{k=1}^m$, we proceed in the same fashion and obtain:

### Theorem 4.9

*Under the assumptions of Theorem 4.8, suppose that the inequalities (4.36) hold. If $L(t)$ satisfies the condition (4.30), then $\dim_H(\mathcal{A}) \le N$, where $N$ is such that*

$$-\overline{h}_1(t)c_1 N^\gamma + \sum_{k=1}^m \overline{h}_{s_k}(t)c_{s_k} N^{s_k} < 0 , \, \gamma > 0 , \, s_k \le 0 , \quad (4.42)$$

*for some $t > 0$. If $N$ is such that*

$$-\frac{1}{1+\gamma}\overline{h}_1(t)c_1 N^\gamma + \sum_{k=1}^m \overline{h}_{s_k}(t)c_{s_k} N^{s_k} < 0 ,$$

$$\gamma > 0 , \, s_k \le 0 \qquad (4.43)$$

*for some $t > 0$, then $\dim_f(\mathcal{A}) \le N$.*

*If $L(t)$ satisfies the condition (4.34), then $\dim_H(\mathcal{A}) \le \dim_f(\mathcal{A}) \le N$, where $N$ is such that*

$$-\overline{h}_0(t) + \sum_{k=1}^m \overline{h}_{s_k}(t)c_{s_k} N^{s_k} < 0 , \quad s_k < 0 \qquad (4.44)$$

*for some $t > 0$.*

Usually in some problems of mathematical physics (e.g., for the Navier–Stokes equations) the functions $\overline{h}_i(t) \equiv \frac{1}{t}\int_0^t h_i(\tau)d\tau$ decrease when $t \to +\infty$, and $h_1(t)$ in (4.30) or $h_0(t)$ in (4.34) does not depend on $t$. In this case we may put $\overline{h}_{s_k}(\infty)$ instead of $\overline{h}_{s_k}(t)$ in (4.41)–(4.44).

# Part II.
# Semigroups generated by evolution equations

# 5

---

# Introduction to Part II

In Part II we consider abstract semi-linear evolution equations mainly of hyperbolic type. They generate semigroups of class $A\mathcal{K}$. Evolution equations of parabolic type generate semigroups of class $\mathcal{K}$. We devote to them only the short Chapter 6, as semi-linear parabolic equations are expounded in a comparatively complete literature. Many publications are devoted to the Navier–Stokes equations which generate (in the two-dimensional case) semigroups of class $\mathcal{K}$. In the first publication on this subject [1] the set $\mathcal{M}$ of all limit states or, which is the same, the minimal global $B$-attractor was found; among its properties the most interesting one is a finiteness of the dynamics $\{V_t\}$ on $\mathcal{M}$ (or alternatively, the finiteness of the number $N_1$ of determining modes). Here we give some comparatively recent results [11] concerning majorants for the number $N_1$ and for the fractal dimension of invariant bounded sets which are better (for small viscosity $\nu$) than before. On the other hand, quasi-linear parabolic equations of general form also generate semigroups of class $\mathcal{K}$, but the presentation of this material requires a separate publication. For this purpose we need to use results on the global unique solvability of boundary value problems for these equations and estimates of a local type. They may be found in the monograph [15]; more recent results are described in the survey [16], which also contains a list of publications. It is, nevertheless, necessary to put this subject in the framework of the theory of semigroups choosing the phase spaces correctly.

We describe now some known general facts used in the next chapters.

Let $H$ be a complete separable Hilbert space, $(\cdot, \cdot)$ and $\|\cdot\|$ denote the inner product and norm in $H$, and $A$ a linear unbounded operator with domain $\mathcal{D}(A)$ dense in $H$. Moreover, $A$ is self-adjoint, positive definite and its inverse $A^{-1}$ is completely continuous. Let us denote by $0 < \lambda_1 \leq \lambda_2 \leq \ldots$ the eigenvalues of $A$ and by $\phi_1, \phi_2, \ldots$ the corresponding orthonormalized eigenelements. Starting from $H$ and $A$ we construct by the usual procedure the space-scale $H_s(A)$, $s \in \mathbf{R}$ : $H_s(A)$ is the domain $\mathcal{D}(A^{s/2})$ of $A^{s/2}$ with the inner product

$$(u, v)_{s,A} = (A^{s/2}u, A^{s/2}v) \tag{5.1}$$

and the norm $\|\cdot\|_{s,A}$. Clearly, $H_0(A) \equiv H$. The spaces $H_s(A)$ and $H_{-s}(A)$ are dual with respect to $H$. We simply write:

$$(u, v) = (A^{s/2}u, A^{-s/2}v) \tag{5.2$_1$}$$

if $u \in H_s(A)$ and $v \in H_{-s}(A)$, and

$$
\begin{aligned}
(u, v)_{r,A} &= (A^{s/2}u, A^{-s/2}v)_{r,A} \\
&= (A^{(r+s)/2}u, A^{(r-s)/2}v)
\end{aligned} \tag{5.2$_2$}
$$

if $u \in H_{s+r}(A)$ and $v \in H_{-s+r}(A)$. It is easy to verify that

$$
\begin{aligned}
&\|u\|_{s,A}^2 \geq \lambda_1^\delta \|u\|_{s-\delta,A}^2 \quad \text{for} \quad \delta > 0, \ u \in H_s(A), \\
&A^{\delta/2}(H_{s-\delta}(A)) = H_s(A), \quad \|A^{\delta/2}u\|_{s,A} = \|u\|_{s+\delta,A}, \\
&(u, v)_{s+\delta,A} = (A^\delta u, v)_{s,A}, \\
&|(u, v)_{s,A}| \leq \|u\|_{s+\delta,A}\|v\|_{s-\delta,A}, \quad \text{etc.}
\end{aligned} \tag{5.3}
$$

In Chapter 6 we use only the scale $H_s(A)$, $s \in \mathbf{R}$, with a fixed operator $A$ and omit $A$ in the notations $H_s(A)$, $(\cdot, \cdot)_{s,A}$ and $\|\cdot\|_{s,A}$. But the dependence on $A$ will be explicitly indicated in Chapter 7, where we deal with operators depending on a parameter. We shall denote the set of all linear bounded operators acting from the Banach space $X$ to the Banach space $Y$ by the symbol $L(X \to Y)$, their norms by the symbol $\|\cdot\|_{L(X \to Y)}$.

# 6

# Estimates for the number of determining modes and the fractal dimension of bounded invariant sets for the Navier-Stokes equations.

It is well known that the Navier–Stokes equations with some boundary conditions can be considered as the equation

$$\partial_t v(t) + \nu A v(t) + f(v(t)) = h \tag{6.1}$$

in a Hilbert space $H$. In the case of homogeneous sticking boundary conditions (i.e. $v(x,t)|_{x \in \partial\Omega} = 0$, where $\Omega$ is a bounded domain in $\mathbf{R}^m$, $m = 2$ or 3, with a smooth boundary $\partial\Omega$), $H \equiv \overset{\circ}{\mathcal{J}}(\Omega)$ is a subspace of the vector-space $L_2^m(\Omega)$. In the case of periodic boundary conditions (here $\Omega$ is a parallelepiped) $H$ is another subspace of $L_2^m(\Omega)$. In both cases $A$ enjoys the properties indicated in Chapter 5 (like the scalar Laplace operator $(-\Delta)$ in $L_2(\Omega)$ with the boundary condition $u|_{\partial\Omega} = 0$). Elements $v$ of $H$ are vector functions $v : x \in \Omega \rightarrow v(x) = (v_1(x), \ldots, v_m(x)) \in \mathbf{R}^m$ and $f(v) = P\mathcal{F}(v)$ where $P$ is the orthoprojector of $L_2^m(\Omega)$ onto $H$ and $\mathcal{F}$ is defined by the mapping:

$$\mathcal{F}(v) : v \rightarrow \sum_{k=1}^{m} v_k(x) \frac{\partial}{\partial x_k} v(x) \in \mathbf{R}^m, \quad x \in \Omega.$$

In [1] it is proved that in the case $m = 2$ (for two-dimensional $\Omega$) the solution operators $V_t$ of problem (6.1) with $h \in H$ form a continuous semigroup $\{V_t, t \in \mathbf{R}^+, H\}$ of class $\mathcal{K}$ and this semigroup has a compact connected minimal global $B$-attractor $\mathcal{M}$, lying in the ball $B_{R_0} = \{v : \|v\| \leq R_0 = \|h\| (\nu\lambda_1)^{-1}\}$. The set $\mathcal{M}$ is bounded in the space $H_2$. It is also maximal among all bounded invariant sets of our semigroup.

The semigroup can be extended on $\mathcal{M}$ to a continuous group $\{V_t, t \in \mathbf{R}^+, \mathcal{M}\}$, and, what is more, this group is in some sense finite-dimensional. It means that there is an integer $N_1$ such that the projection $P^{N_1}\gamma(v)$ of any complete trajectory $\gamma(v)$, $v \in \mathcal{M}$ in the subspace span $\{\phi_1, \ldots, \phi_{N_1}\} \equiv P^{N_1}H$ determines the trajectory $\gamma(v)$ (here $\phi_k, k = 1, \ldots, N_1$, are eigenelements of $A$). More precisely, if for $\gamma(v) = \{V_t(v), t \in \mathbf{R}\}$ and $\gamma(\tilde{v}) = \{V_t(\tilde{v}), t \in \mathbf{R}\}$ with $v, \tilde{v} \in \mathcal{M}$ we have the equalities $P^{N_1}V_t(v) = P^{N_1}V_t(\tilde{v})$ for all $t \in \mathbf{R}$, then $V_t(v) = V_t(\tilde{v})$ for all $t \in \mathbf{R}$.

We call the smallest of such integer numbers the *number $N_1$ of determining modes* for $\mathcal{M}$ (or for $\mathcal{A}$ if we consider another invariant set $\mathcal{A}$). In [1] a majorant for $N_1$ is computed. Here we shall deduce a majorant which is better for small $\nu^{-1}$. Let us remark that we do not use the fact that $\mathcal{M}$ is an attractor.

For an arbitrary complete trajectory $v(t)$, $t \in \mathbf{R}$ lying in $\mathcal{M}$ we have the estimates:

$$\sup_{t \in R} \|v(t)\| \leq R_0 \equiv \|h\| (\nu\lambda_1)^{-1} \qquad (6.2)$$

and

$$\nu \int_\tau^t \|v(\xi)\|_1^2 \, d\xi \leq \frac{1}{2}R_0^2 + \|h\|^2 (\lambda_1\nu)^{-1}|t - \tau|, \qquad (6.3)$$

where $\lambda_k$, $k = 1, 2, \ldots$, are the eigenvalues of $A$. The difference $u(t) = v(t) - \tilde{v}(t)$ of two solutions of (6.1) satisfies the equality:

$$\partial_t u(t) + \nu A u(t) + P(\tilde{v}_k(t)u_{x_k}(t) + u_k(t)v_{x_k}(t)) = 0, t \in \mathbf{R}. \quad (6.4)$$

From (6.4) we deduce

$$\frac{1}{2}\frac{d}{dt}\|u(t)\|^2 + \nu\|u(t)\|_1^2 = -\int_\Omega u_k(t, x)v_{x_k}(t, x)u(t, x)dx$$

$$\leq \|v(t)\|_1 \|u(t)\|_{L_4(\Omega)}^2 \leq \frac{1}{\sqrt{2}}\|v(t)\|_1 \|u(t)\|_1 \|u(t)\|$$

$$\leq \frac{\nu}{2}\|u(t)\|_1^2 + \frac{1}{4\nu}\|v(t)\|_1^2\|u(t)\|^2,$$

and from this inequality it follows that

$$\frac{d}{dt}\|u(t)\|^2 + \nu\|u(t)\|_1^2 \le (2\nu)^{-1}\|v(t)\|_1^2\| u(t)\|^2 . \tag{6.5}$$

Suppose that $P^N u(t) = 0$ for all $t \in \mathbf{R}$. Then $u(t) = Q^N u(t)$ where $Q^N := I - P^N$ and

$$\|u(t)\|_1^2 \ge \lambda_{N+1}\|u(t)\|^2 \tag{6.6}$$

(as well as for an arbitrary element of $Q^N H$), so that

$$\frac{d}{dt}\|u(t)\|^2 \le -h(t)\|u(t)\|^2 , \tag{6.7}$$

with $h(t) = \nu\lambda_{N+1} - (2\nu)^{-1}\|v(t)\|_1^2$. If $\int_\tau^t h(\xi)d\xi \to +\infty$ when $\tau \to -\infty$ then $u(t) = 0$ for all $t \in \mathbf{R}$. By (6.3) this will hold if

$$\lambda_{N+1} > \nu^{-4}\|h\|^2(2\lambda_1)^{-1} . \tag{6.8}$$

As $\lambda_k = O(k)$ we can satisfy (6.8) by choosing an $N$ which satisfies the inequality

$$N \le c_1\nu^{-4} + c_1', \text{ with some } c_1, c_1' \in \mathbf{R}^+ . \tag{6.9}$$

In the case of periodic boundary conditions we have the following estimates for all $v(t) \in \mathcal{M}$:

$$\sup_{t \in R} \|v(t)\| \le c\nu^{-1}, \ \sup_{t \in R} \|v(t)\|_1 \le c\nu^{-1} , \tag{6.10}$$

$$\frac{1}{t - \tau} \int_\tau^t \|v(\xi)\|_2^2 \, d\xi \le c\nu^{-2}[1 + \nu^{-1}(t - \tau)^{-1}] , \tag{6.11}$$

$$-\infty < \tau < t < \infty ,$$

which are better than (6.3) for large $\nu^{-1}$.

Using these estimates we can satisfy the requirement (6.8) if $N$ satisfies the inequality:

$$N \le c_2\nu^{-2}|\ln\frac{1}{\nu}| + c_2', \text{ with some } c_2, c_2' \in \mathbf{R}^+ . \tag{6.12}$$

In the case $m = 3$ (for three-dimensional $\Omega$) we also can estimate the number of determining modes for any invariant set $\mathcal{A}$ bounded in $H_1$, following the same procedure. For both boundary conditions (for sticking and periodic) the estimates have the form

$$N \le c_3\nu^{-9}m_1^3 + c_3', \quad c_3, c_3' \in \mathbf{R}^+ , \tag{6.13}$$

where $m_1 \equiv \sup_{v \in \mathcal{A}} \|v\|_1$.

So the following Theorem holds:

**Theorem 6.1**

*The number $N_1$ of determining modes for $\mathcal{M}$, in the case*

*$m = 2$ and sticking boundary conditions, has the majorant indicated in (6.9) and for periodic boundary conditions the one in (6.12). In the case $m = 3$ with either sticking or periodic conditions the majorants of number $N_1$ for invariant sets $\mathcal{A}$ bounded in $H_1$ have the form (6.13).*

### Majorants of fractal dimensions

Let us denote by $d_f^{(s)}(\mathcal{A})$ the fractal dimension of a compact set $\mathcal{A}$ as a subset of space $H_s$. Generally the finiteness of $d_f^{(s+\epsilon)}(\mathcal{A}), \epsilon > 0$, for a compact $\mathcal{A}$ in the space $H_{s+\epsilon}$, does not follow from the finiteness of $d_f^{(s)}(\mathcal{A})$. But for the Navier–Stokes equations and for some other partial differential equations it is possible to evaluate $d_f^{(s)}(\mathcal{A})$ for any $s$ using theorems from Chapter 4. For the Navier–Stokes equations we have computed the majorants for $d_f^{(k)}(\mathcal{M})$, $k = 0, 1$, in the case $m = 2$ and for $d_f^{(k)}(\mathcal{A})$, $k = 0, 1$, in the case $m = 3$. Namely

### Theorem 6.2

*The numbers $d_f^{(k)}(\mathcal{M})$, $k = 0, 1$, for the two-dimensional Navier–Stokes equations with sticking boundary conditions, have the following majorants:*

$$d_f^{(0)}(\mathcal{M}) \leq c_4 \nu^{-4} + c_4', \tag{6.14}$$

$$d_f^{(1)}(\mathcal{M}) \leq c_5 \nu^{-2} m_1^2 [(\ln \frac{1}{\nu})^2 + (\ln m_1)^{1/2}] + c_5', \tag{6.15}$$

*where $m_1 = \sup_{v \in \mathcal{M}} \|v\|_1$. For periodic boundary conditions, we have*

$$d_f^{(k)}(\mathcal{M}) \leq c_6 \nu^{-2} |\ln \frac{1}{\nu}| + c_6', \quad k = 0, 1. \tag{6.16}$$

*The majorants of $d_f^{(k)}(\mathcal{A})$ for the three-dimensional Navier–Stokes equations with sticking or periodic boundary conditions have the form*

$$d_f^{(k)}(\mathcal{A}) \leq c_7 \nu^{-9} m_1^3 + c_7', \quad k = 0, 1, \tag{6.17}$$

*where $m_1 = \sup_{v \in A} \|v\|_1$.*

*The constants $c_\ell, c_\ell'$ are determined by the norm $\|h\|$ of the free term (the forces) $h$ in the Navier–Stokes equations and in (6.15), (6.16) by the norm $\|h\|_1$. In addition, they also depend on $\Omega$.*

# 7

---

# Evolution equations of hyperbolic type

## 7.1 Introduction

We now investigate semigroups originated by the problem

$$\partial_{tt}^2 v(t) + \nu \partial_t v(t) + A v(t) + f(v(t)) = h \,,$$

$$v|_{t=0} = \Psi_0, \quad \partial_t v|_{t=0} = \Psi_1 \,. \tag{7.1}$$

Here $\nu = \text{constant} > 0$, $A$ is a linear unbounded self-adjoint positively defined operator in the Hilbert space $H$, which has a completely continuous inverse operator; $h$ is a fixed element of $H$, $f$ is a certain nonlinear (generally unbounded) operator. Below we shall formulate conditions for $f$ under which the problem (7.1) is globally and uniquely solvable and a semigroup it generates possesses some properties enabling us to find its minimal attractors. By using the procedure of Chapter 5, for an operator $A$ defined in a dense domain, we introduce a space-scale $H_s(A)$, $s \in \mathbf{R}$. To apply the results of Part I, we shall reformulate the problem (7.1) as the Cauchy problem for first order equations.

Usually the vector function $\vec{v}(t) = \begin{pmatrix} v_0(t) \\ v_1(t) \end{pmatrix}$ is chosen to be unknown, where $v_0$ is the solution $v(t)$ of the problem (7.1) and $v_1(t) = \partial_t v(t)$.

In this variant the problem (7.1) acquires the form

$$\partial_t v_0(t) = v_1(t) \,,$$

$$\partial_t v_1(t) = -A v_0(t) - \nu v_1(t) - f(v_0(t)) + h \,, \tag{7.2}$$

$$v_0|_{t=0} = \Psi_0 \,, \quad v_1|_{t=0} = \Psi_1$$

or, what is the same,

$$\partial_t \vec{v}(t) = a\vec{v}(t) + \vec{f}(\vec{v}(t)) + \vec{h} ;$$
$$\vec{v}|_{t=0} = \vec{\Psi} ,$$

(7.3)

where

$$a = \begin{pmatrix} 0 & I \\ -A & -\nu I \end{pmatrix} , \quad \vec{f}(\vec{v}(t)) = \begin{pmatrix} 0 \\ -f(v_0(t)) \end{pmatrix} ,$$
$$\vec{h} = \begin{pmatrix} 0 \\ h \end{pmatrix} , \quad \vec{\Psi} = \begin{pmatrix} \Psi_0 \\ \Psi_1 \end{pmatrix} .$$

However, for establishing some important properties of semigroup generated by problem (7.1) and associated problems, we use another formulation. That is we introduce the vector-function $\vec{v}(t, \alpha)$ defined by $\vec{v}(t)$ with the help of the equality

$$\vec{v}(t, \alpha) = C(\alpha)\vec{v}(t), \quad C(\alpha) = \begin{pmatrix} 1 & 0 \\ \alpha & 1 \end{pmatrix} ,$$

where $\alpha$ is a positive number subjected to certain restrictions. The vector function $\vec{v}(t, \alpha)$ is connected with the solution $v(t)$ of (7.1) by the equalities:

$$v_0(t, \alpha) = v(t), \quad v_1(t, \alpha) = \alpha v(t) + \partial_t v(t) .$$

It is easy to prove that, if $\vec{v}(t)$ is a solution of (7.3), then $\vec{v}(t, \alpha)$ is the solution of the problem

$$\partial_t \vec{v}(t, \alpha) = a(\alpha)\vec{v}(t, \alpha) + \vec{f}(\vec{v}(t, \alpha)) + \vec{h} ,$$
$$\vec{v}|_{t=0} = \vec{\Psi}(\alpha) ,$$

(7.4)

where

$$a(v) = \begin{pmatrix} -\alpha I & I \\ -A(\alpha) & -(\nu - \alpha)I \end{pmatrix} ,$$
$$A(\alpha) = A - \alpha(\nu - \alpha)I , \quad \vec{\Psi}(\alpha) = \begin{pmatrix} \Psi_0 \\ \alpha \Psi_0 + \Psi_1 \end{pmatrix} ;$$

and vice versa, if $\vec{v}(t, \alpha)$ is a solution of the problem (7.4), then $\vec{v}(t) = C^{-1}(\alpha)\vec{v}(t, \alpha) = C(-\alpha)\vec{v}(t, \alpha)$ is a solution of the problem (7.3).

In terms of the components $v_0(t, \alpha)$, $v_1(t, \alpha)$ of $\vec{v}(t, \alpha)$, the system (7.4) looks like:

$$\partial_t v_0(t, \alpha) = -\alpha v_0(t, \alpha) + v_1(t, \alpha) ,$$
$$\partial_t v_1(t, \alpha) = -A(\alpha)v_0(t, \alpha) - (\nu - \alpha)v_1(t, \alpha)$$
$$\qquad - f(v_0(t, \alpha)) + h ,$$
$$v_0|_{t=0} = \Psi_0 \equiv \Psi_0(\alpha) , \quad v_1|_{t=0} = \alpha \Psi_0 + \Psi_1 \equiv \Psi_1(\alpha) .$$

(7.5)

It is clear that for $\alpha = 0$, the problems (7.4) and (7.3) coincide and $a(0) = a$, $\vec{\Psi}(0) = \vec{\Psi}$.

The parameter $\alpha$ will satisfy the following conditions

$$\alpha \in [0, \tfrac{\nu}{2}], \quad \frac{\lambda_1(A(\alpha))}{\lambda_1(A)} = 1 - \frac{\alpha(\nu - \alpha)}{\lambda_1(A)} \equiv m^2(\alpha) > 0. \quad (7.6)$$

Here $\lambda_1(A(\alpha)) \equiv \lambda_1(\alpha)$ is the first eigenvalue of the operator $A(\alpha)$. As before, we shall denote the complete orthonormalized system of the eigenelements of the operator $A$ by $\{\phi_k\}_{k=1}^\infty$ and the corresponding eigenvalues by $\{\lambda_k\}_{k=1}^\infty$. The eigenvalues of the operator $A(\alpha)$ are equal to $\lambda_k(A(\alpha)) = \lambda_k(A) - \alpha(\nu - \alpha) \equiv \lambda_k(\alpha)$ and the eigenfunctions of $A(\alpha)$ and $A$ are the same. We shall use the space-scales $H_s(A(\alpha)) \equiv H_{s,\alpha}$, $s \in \mathbf{R}$, constructed with the help of the operator $A(\alpha)$ as described in Chapter 5. The scalar product in $H_{s,\alpha}$ will be denoted by the symbol $(\cdot, \cdot)_{s,\alpha}$ and the norm by $\|\cdot\|_{s,\alpha}$. When $\alpha = 0$, instead of $H_{s,0}$, $(\cdot, \cdot)_{s,0}$ and $\|\cdot\|_{s,0}$ we shall use $H_s$, $(\cdot, \cdot)_s$ and $\|\cdot\|_s$ (in accordance with Chapter 6).

Under the conditions (7.6), $H_{s,\alpha}$ and $H_s$ coincide as sets and their norms are equivalent. In fact, elementary calculations give the following inequalities

$$m^s(\alpha)\|u\|_s \leq \|u\|_{s,\alpha} \equiv \|A^{s/2}(\alpha)u\|$$
$$\leq \|A^{s/2}u\| \equiv \|u\|_s \quad \text{for } s \geq 0, \quad (7.7)$$
$$\|u\|_s \leq \|u\|_{s,\alpha} \leq m^s(\alpha)\|u\|_s \quad \text{for } s \leq 0,$$

where $\|\cdot\|$, as usual, is the norm in $H$.

As the phase space for problem (7.4) we shall take the space $X_{s,\alpha} \equiv H_{s+1,\alpha} \times H_{s,\alpha}$. The scalar product in $X_{s,\alpha}$ is defined by

$$\begin{aligned}(u, v)_{X_{s,\alpha}} &= (u_0, v_0)_{s+1,\alpha} + (u_1, v_1)_{s,\alpha} \\ &\equiv (A^{(s+1)/2}(\alpha)u_0, A^{(s+1)/2}(\alpha)v_0) \\ &\quad + (A^{s/2}(\alpha)u_1, A^{s/2}(\alpha)v_1),\end{aligned} \quad (7.8)$$

where $(\cdot, \cdot)$, as usual, is the scalar product in $H$. The norm in $X_{s,\alpha}$ will be denoted by the symbol $\|\cdot\|_{X_{s,\alpha}}$; when $\alpha = 0$ we shall use the notation $X_s$, $(\cdot, \cdot)_{X_s}$ and $\|\cdot\|_{X_s}$.

From (7.7) it follows that for an arbitrary $\vec{u} \in X_s$ (or, what is the same, $\vec{u} \in X_{s,\alpha}$)

$$m^{s+1}(\alpha)\|\vec{u}\|_{X_s} \leq \|\vec{u}\|_{X_{s,\alpha}} \leq \|\vec{u}\|_{X_s}, \quad s \geq 0,$$
$$m^{s+1}(\alpha)\|\vec{u}\|_{X_s} \leq \|\vec{u}\|_{X_{s,\alpha}} \leq m^s(\alpha)\|\vec{u}\|_{X_s}, \quad s \in [-1, 0], \quad (7.9)$$
$$\|\vec{u}\|_{X_s} \leq \|\vec{u}\|_{X_{s,\alpha}} \leq m^s(\alpha)\|\vec{u}\|_{X_s}, \quad s \leq -1,$$

The same space-scale $X_{s,\alpha}$, $s \in \mathbf{R}$, is constructed from the space $X_{0,\alpha} \equiv H_1(A(\alpha)) \times H$ by the standard procedure described in Chapter 5, with the help of the unbounded operator $\vec{A}(\alpha) \equiv \begin{pmatrix} A(\alpha) & 0 \\ 0 & A(\alpha) \end{pmatrix}$:

$$\mathcal{D}(\vec{A}(\alpha)) \equiv \mathcal{D}(A^{3/2}(\alpha)) \times \mathcal{D}(A(\alpha)) \subset X_0 \to X_0 \,.$$

The latter enjoys all the properties required by this procedure. The spectrum of $\vec{A}(\alpha)$ consists of the numbers $\{\lambda_k^{\pm} = \lambda_k(\alpha)\}_{k=1}^{\infty}$, each $\lambda_k(\alpha)$ corresponding to two linearly independent eigenvectors, normalized in $X_0$:

$$\vec{\Psi}_k^+(\alpha) = \begin{pmatrix} \lambda_k^{-1/2}(\alpha)\phi_k \\ 0 \end{pmatrix} \quad \text{and} \quad \vec{\Psi}_k^-(\alpha) = \begin{pmatrix} 0 \\ \phi_k \end{pmatrix} \,.$$

The operator $\vec{A}^{1/2}(\alpha)$ establishes a one-to-one correspondence between $X_{s+1,\alpha}$ and $X_{s,\alpha}$, and $\|\vec{A}^{1/2}\vec{u}\|_{X_{s,\alpha}} = \|\vec{u}\|_{X_{s+1,\alpha}}$.

Let us introduce one more notation,

$$a_0(\alpha) \equiv \begin{pmatrix} 0 & I \\ -A(\alpha) & 0 \end{pmatrix} \,,$$

which is the principal part of the operator $a(\alpha)$. It is easy to verify that $a_0(\alpha)$ also gives a one-to-one correspondence between $X_{s+1,\alpha}$ and $X_{s,\alpha}$, and $\|a_0(\alpha)\vec{u}\|_{X_{s,\alpha}} = \|\vec{u}\|_{X_{s+1,\alpha}}$. Moreover

$$(a_0(\alpha)\vec{u}, \vec{v})_{X_{s,\alpha}} = -(\vec{u}, a_0(\alpha)\vec{v})_{X_{s,\alpha}} \tag{7.10}$$

for arbitrary $\vec{u}, \vec{v} \in X_{s+1,\alpha}$.

We begin our analysis of the problem (7.1), with the investigation of some linear problems.

### 7.2 Linear problems

The problem

$$\partial_t \vec{u}(t) = a\vec{u}(t) + \vec{g}(t) \,, \quad \vec{u}|_{t=0} = \vec{\Psi} \tag{7.11}$$

is the object of our attention. Here $a = \begin{pmatrix} 0 & I \\ -A & -\nu I \end{pmatrix}$ is the same as in the preceding section, $\vec{g}(.) = \begin{pmatrix} 0 \\ g(.) \end{pmatrix}$ is a fixed element of $L_{p,loc}(\mathbf{R}, X_s)$ with $p \geq 1$ and $s \in \mathbf{R}$, and $\vec{u}(t) = \begin{pmatrix} u_0(t) \\ u_1(t) \end{pmatrix}$ is function we seek. We want to prove the solvability of (7.11) in the space $X_s$ for an arbitrary $\vec{\Psi} \in X_s$. The main energy relation for the problem is

$$\frac{1}{2}\frac{d}{dt}\|\vec{u}(t)\|_{X_s}^2 = -\nu\|u_1(t)\|_s^2 + (g(t), u_1(t))_s \,, \tag{7.12}$$

which follows from (7.11) multiplied in $X_s$ by $\vec{u}(t)$ and from the property (7.10) when $\alpha = 0$. By using (7.12) it is easy to estimate $\|u(t)\|_{X_s}$ through $\|\vec{\Psi}\|_{x_s}$ and $|\int_0^t \|g(\tau)\|_s d\tau|$. But it does not reflect the important property of the solutions of the problem (7.11): that is, the exponential decay of $\|\vec{u}(t)\|_{X_s}$, when $t \to +\infty$, of solutions of the homogeneous equation (7.11) (i.e. when $g(t) \equiv 0$) and the boundedness of $\|\vec{u}(t)\|_{X_s}$ on the semi-axis $t \in \mathbf{R}^+$ in the general case if $\sup_{t \in \mathbf{R}} \|g(t)\|_s < +\infty$. This can be proved by different means. One of them is to develop the solution $\vec{u}(t)$ in a Fourier series by the eigenelements of the operator $a$. The expansion enables one to know the behaviour of $\|\vec{u}(t)\|_{X_s}$ when $t \to \infty$. However we would prefer a different procedure, keeping in mind the further applications of the results of Part I to the problem (7.3). As in section 7.1, we introduce the functions

$$\vec{u}(t,\alpha) \equiv C(\alpha)\vec{u}(t), \ C(\alpha) = \begin{pmatrix} 1 & 0 \\ \alpha & 1 \end{pmatrix}, \tag{7.13}$$

where $\vec{u}(t)$ are solutions of (7.11) and $\alpha$ is a number obeying the inequalities (7.6). The $\vec{u}(t,\alpha)$ are solutions of the following problem:

$$\partial_t \vec{u}(t,\alpha) = a(\alpha)\vec{u}(t,\alpha) + \vec{g}(t),$$

$$\vec{u}|_{t=0} = \vec{\Psi}(\alpha) = \begin{pmatrix} \Psi_0 \\ \alpha\Psi_0 + \Psi_1 \end{pmatrix} \tag{7.14}$$

where $a(\alpha) = \begin{pmatrix} -\alpha I & I \\ -A(\alpha) & -(\nu - \alpha)I \end{pmatrix}$. In coordinates (7.14) looks like

$$\partial_t u_0(t,\alpha) = -\alpha u_0(t,\alpha) + u_1(t,\alpha),$$

$$\partial_t u_1(t,\alpha) = -A(\alpha)u_0(t,\alpha) - (\nu - \alpha)u_1(t,\alpha) + g(t), \tag{7.15}$$

$$u_0|_{t=0} = \Psi_0(\alpha) = \Psi_0, \ u_1|_{t=0} = \Psi_1(\alpha) = \alpha\Psi_0 + \Psi_1.$$

The following relation holds:

$$\frac{1}{2}\frac{d}{dt}\|u(t,\alpha)\|^2_{X_{s,\alpha}}$$
$$= -\alpha\|u_0(t,\alpha)\|^2_{s+1,\alpha} - (\nu - \alpha)\|u_1(t,\alpha)\|^2_{s,\alpha} \tag{7.16}$$
$$+ (g(t), u_1(t,\alpha))_{s,\alpha}$$

which is the analogue of (7.12) for (7.14). It is the result of the multiplication of (7.14) in $X_{s,\alpha}$ by $\vec{u}(t,\alpha)$, bearing in mind property

(7.10). The inequalities

$$- \nu \|\vec{u}(t, \alpha)\|_{X_{s,\alpha}} - \|\vec{g}(t)\|_{s,\alpha}$$

$$\leq \frac{\mathrm{d}}{\mathrm{d}t} \|\vec{u}(t, \alpha)\|_{X_{s,\alpha}} \tag{7.17}$$

$$\leq - \alpha \|\vec{u}(t, \alpha)\|_{X_{s,\alpha}} + \|\vec{g}(t)\|_{s,\alpha}$$

follow from (7.16), because we assumed that $\alpha \in [0, \nu/2]$, while from (7.17) we may deduce the following estimates

$$\|\vec{u}(t, \alpha)\|_{X_{s,\alpha}} \leq e^{-\alpha t} \|\vec{u}(0, \alpha)\|_{X_{s,\alpha}}$$

$$+ \int_0^t e^{-\alpha(t-\tau)} \|g(\tau)\|_{s,\alpha} \mathrm{d}\tau, t \in \mathbf{R}^+ , \tag{7.18$_1$}$$

$$\|\vec{u}(t, \alpha)\|_{X_{s,\alpha}} \leq e^{-\nu t} \|\vec{u}(0, \alpha)\|_{X_{s,\alpha}}$$

$$+ \int_t^0 e^{-\nu(t-\tau)} \|g(\tau)\|_{s,\alpha} \mathrm{d}\tau, t \in \mathbf{R}^- . \tag{7.18$_2$}$$

These inequalities allow us to obtain estimates of the same kind for $\|\vec{u}(t)\|_{X_s}$. In fact, for an arbitrary $u \in X_s$ and $\vec{u}(\alpha) \equiv C(\alpha)\vec{u}$, where $C(\alpha)$ is defined in (7.13), the inequalities

$$\|\vec{u}(\alpha)\|_{X_s} \leq \mu_1(\alpha)\|\vec{u}\|_{X_s} \leq \mu_1^2(\alpha)\|\vec{u}(\alpha)\|_{X_s} , \tag{7.19$_1$}$$

where $\mu_1(\alpha) = \max\left\{\sqrt{1+\alpha}, \sqrt{1 + \alpha\lambda_1^{-1} + \alpha^2\lambda_1^{-1}}\right\}$ are confirmed routinely. From (7.19$_1$) and (7.9) we deduce the inequalities

$$\|\vec{u}\|_{X_s} \leq m_1(s, \alpha)\|\vec{u}(\alpha)\|_{X_{s,\alpha}} ,$$

$$\|\vec{u}(\alpha)\|_{X_{s,\alpha}} \leq m_2(s, \alpha)\|\vec{u}\|_{X_s} , \tag{7.19$_2$}$$

where

$$m_1(s, \alpha) = \begin{cases} \mu_1(\alpha)m^{-(s+1)}(\alpha), & \text{for } s \geq -1 \\ \mu_1(\alpha), & \text{for } s \leq -1 \end{cases}$$

$$m_2(s, \alpha) = \begin{cases} \mu_1(\alpha), & \text{for } s \geq 0 \\ \mu_1(\alpha)m^s(\alpha), & \text{for } s \leq 0 \end{cases}$$

For $\alpha$ satisfying requirements (7.6), because of (7.19$_2$) and (7.7) we obtain from (7.18$_k$) the main a priori estimates:

$$\|\vec{u}(t)\|_{X_s} \leq m_3(s, \alpha)e^{-\alpha t}\|\vec{u}(0)\|_{X_s}$$

$$+ m_4(s, \alpha) \int_0^t e^{-\alpha(t-\tau)} \|g(\tau)\|_s \mathrm{d}\tau , \tag{7.20$_1$}$$

$$\|\vec{u}(t)\|_{X_s} \leq m_3(s, \alpha)e^{-\nu t}\|\vec{u}(0)\|_{X_s}$$

$$+ m_4(s, \alpha) \int_t^0 e^{-\nu(t-\tau)} \|g(\tau)\|_s \mathrm{d}\tau , \tag{7.20$_2$}$$

for $t \in \mathbf{R}^+$ and $t \in \mathbf{R}^-$ respectively, where

$$m_3(s, \alpha) = m_1(s, \alpha) m_2(s, \alpha)$$

and

$$m_4(s, \alpha) = m_1(s, \alpha) \text{if } s \geq 0\,,$$

$$m_4(s, \alpha) = m_1(s, \alpha) m^s(\alpha) \quad \text{if } s \leq 0$$

for solutions of the problem (7.11). The explicit dependence of $m_k(\cdot)$ on $s$ and on $\alpha$ is not important for us. The essential things are:

(a) the requirements (7.6) enable us to choose $\alpha$ positive and deduce that, for an arbitrary $s \in \mathbf{R}$, $\|\vec{u}(t)\|_{X_s}$ tends to zero when $t \to +\infty$.

(b) when $\alpha$ tends to zero, $\mu_1(\alpha)$, $m(\alpha)$ and $m_k(s, \alpha)$ tend to 1.

Now let us go on to prove the unique solvability of the problem (7.11).

We shall call a *weak solution* of the problem (7.11) a function $\vec{u} \colon \mathbf{R} \to X_r$ for which all projections $(u_0(t), \phi_k)$, $(u_1(t), \phi_k)$, $k = 1, 2, \dots$ are absolutely continuous functions of $t \in \mathbf{R}$ and, for almost any $t$, satisfy the following equalities:

$$\partial_t(u_0(t), \phi_k) = (u_1(t), \phi_k)\,;$$
$$\partial_t(u_1(t), \phi_k) = -\lambda_k(u_0(t), \phi_k) - \nu(u_1(t), \phi_k)$$
$$+ (g(t), \phi_k)\,; \qquad (7.21)$$
$$(u_0, \phi_k)|_{t=0} = (\Phi_0, \phi_k)\,,$$
$$(u_1, \phi_k)|_{t=0} = (\Phi_1, \phi_k)\,, \quad k = 1, 2, \dots.$$

In this terminology the number $r$ is deliberately ignored despite the fact it appears in the condition $\vec{u} \colon \mathbf{R} \to X_r$. Its value is insignificant, only the fact that $r > -\infty$ being important. It should be remembered that for the elements $u \in H_s$ and $v \in H_{-s}$, $(u, v)$ is the number $(A^{s/2}u, A^{-s/2}v)$ — the inner product in $H$ of elements $A^{s/2}u \in H$ and $A^{-s/2}v \in H$. This is the meaning of the brackets $(u_\ell(t), \phi_k)$ in (7.21).

It is easy to prove the following theorem

**Theorem 7.1**
*Problem (7.11) has no more than one weak solution.*
We shall apply this theorem to linear and nonlinear problems as a means of identifying their solutions when we have preliminary and

incomplete information, but when we know that they are weak solutions of a problem of the type (7.11).

Let us prove the following existence theorem:

### Theorem 7.2

*If $g \in L_{p,loc}(\mathbf{R}, H_s)$, $s \in \mathbf{R}$, $p \geq 1$ then problem (7.11) for an arbitrary $\vec{\Psi} \in X_s$ has the unique solution $\vec{u} \in C(\mathbf{R}, X_s)$ with $\partial_t \vec{u} \in L_{p,loc}(\mathbf{R}, X_{s-1})$. For almost all $t$ it satisfies the equation (7.11) (in space $X_{s-1}$) and the energy relation (7.12). For this solution the estimates (7.20$_k$) are true (for all $t$). Any weak solution of problem (7.11) coincides with this $\vec{u}$. If, additionally, $g \in C(\mathbf{R}, H_{s-1})$ then $\partial_t \vec{u} \in C(\mathbf{R}, X_{s-1})$ and equation (7.11) is fulfilled for all $t \in \mathbf{R}$.*

It is easy to prove that all the statements of Theorem 7.2 are true if $\Psi_0 = \sum_{k=1}^{N} a_k \phi_k$, $\Psi_1 = \sum_{k=1}^{N} b_k \phi_k$, and $g(t) = \sum_{k=1}^{N} g_k(t) \phi_k$ with $g_k \in L_{p,loc}(\mathbf{R})$ or $g_k \in C(\mathbf{R})$.

Now let $\vec{\Psi}$ be an arbitrary element of $X_s$ and $g(\cdot)$ an arbitrary element of $L_{p,loc}(\mathbf{R}, H_s)$. The sums of their Fourier-series, $\vec{\Psi}^N$ and $g^N(\cdot)$, containing only $\phi_1, \ldots, \phi_N$, approximate $\vec{\Psi}$ and $g(\cdot)$ when $N \to \infty$ as follows: $\vec{\Psi}^n \to \vec{\Psi}$ in the norm $X_s$, and $g^N(\cdot) \to g(\cdot)$, in the norm $L_p((-T, T), H_s)$, for an arbitrary $T < +\infty$. Because of this and the fact that the estimates (7.20$_k$) hold for the solution $\vec{u}^N(t)$ of the problem (7.11), corresponding to $\vec{\Psi}^N$ and $g^N(\cdot)$, and their differences $\vec{u}^{N_1}(t) - \vec{u}^{N_2}(t)$ for arbitrary $N_1, N_2 < \infty$, the sequence $\{\vec{u}^N(\cdot)\}_{N=1}^{\infty}$ is fundamental in the spaces $C([-T, T], X_s)$. for any $T < +\infty$. Since the spaces $C([-T, T], X_s)$ are complete, the sequence $\{\vec{u}^N(\cdot)\}_{N=1}^{\infty}$ converges in their norms to an element $\vec{u} \in C(\mathbf{R}, X_s)$. Hence $a\vec{u}^N(\cdot)$ converges to $a\vec{u}(\cdot)$ in the norm of $C([-T, T], X_{s-1})$ (one must remember that $a \in L(X_s \to X_{s-1})$). Furthermore, as $\partial_t \vec{u}^N(t) = a\vec{u}^N(t) + \vec{g}^N(t)$ and $\vec{g}^N(\cdot) \to \vec{g}(\cdot)$ in the norms of $L_p([-T, T], X_s)$, $T < +\infty$, then $\partial_t \vec{u}^N(\cdot)$ converges in the norms of $L_p([-T, T], X_{s-1})$ to an element of $L_{p,loc}(\mathbf{R}, X_{s-1})$, which, according to the theory of generalized derivatives, is equal to $\partial_t \vec{u}(\cdot)$. If, additionally, $g \in C(\mathbf{R}, H_{s-1})$, then $\vec{g}^N(\cdot)$ converges to $\vec{g}(\cdot)$ in the norms of $C([-T, T], X_{s-1})$ and $\partial_t \vec{u}^N(\cdot)$ converges to $\partial_t \vec{u}$ in the same norms. In this case equation (7.11) will be fulfilled (in $X_{s-1}$) for all $t$, and in the first case (i.e. for $g \in L_{p,loc}(\mathbf{R}, X_s)$) for almost all $t$. Relation (7.12) is fulfilled (for almost all $t$), since it is true for all $u^N(t)$ with $g(t) = g^N(t)$, and each term of its right hand side converges to the corresponding term of the right hand side of (7.12) in the norms of $L_p((-T,T))$, $T < +\infty$.

The uniqueness follows from Theorem 7.1 since the solution $\vec{u}(t)$ is a weak solution of problem (7.11).

Theorem 7.2 for the case $g(t) \equiv 0$ guarantees the existence of the operators $U_t \in L(X_s \to X_s)$ associating to $\vec{\Psi} \in X_s$ the solution $\vec{u}(t)$ of the problem

$$\partial_t \vec{u}(t) = a\vec{u}(t), \ \vec{u}|_{t=0} = \vec{\Psi}, \tag{7.22}.$$

The inequalities $(7.20_k)$ give the following estimates:

$$\|U_t\|_{L(X_s \to X_s)} \leq m_3(\alpha,s)e^{-\alpha t}, \ t \in \mathbf{R}^+, \tag{7.23_1}$$

$$\|U_t\|_{L(X_s \to X_s)} \leq m_3(\alpha,s)e^{-\nu t}, \ t \in \mathbf{R}^-. \tag{7.23_2}$$

It is obvious that $\{U_t, t \in \mathbf{R}, X_s\}$ is a continuous group of linear bounded operators.

As in the finite-dimensional case, the solution $\vec{u}(t)$ of problem (7.11) may be represented by means of Duhamel's principle as follows

$$\vec{u}(t) = U_t \vec{\Psi} + \int_0^t U_{t-\tau} \vec{g}(\tau) d\tau. \tag{7.24}$$

This can be proved using the above mentioned properties of the operators $U_t$ and the fact that $a \in L(X_s \to X_{s-1})$. But, otherwise, the representation (7.24) is true for the finite-dimensional approximations $\vec{u}^N(t)$ and the estimates $(7.20_k)$ allow us to pass to the limit when $N \to \infty$ and to obtain (7.24) for $\vec{u}(t)$ as the limit of $\vec{u}^N(t)$.

Problem (7.11), as proved above, is connected with the problem:

$$\partial_{tt}^2 u(t) + \nu \partial_t u(t) + Au(t) = g(t),$$
$$u|_{t=0} = \Psi_0, \quad \partial_t u|_{t=0} = \Psi_1 \tag{7.25}$$

in the following way. The first component $u_0(t)$ of the solution $\vec{u}(t)$ is the solution $u(t)$ of problem (7.25) and the second component of $\vec{u}(t)$ is equal to $\partial_t u(t)$. Therefore from Theorem 7.2 for problem (7.25) we get:

### Theorem 7.2′

If $g \in L_{p,loc}(\mathbf{R}, H_s)$, $s \in \mathbf{R}$, $p \geq 1$, then problem (7.25) has the unique solution $u \in C(\mathbf{R}, H_{s+1})$ with $\partial_t u \in C(\mathbf{R}, H_s)$ and $\partial_t^2 u \in L_{p,loc}(\mathbf{R}, H_{s-1})$. The solution $u(t)$ for almost all $t \in \mathbf{R}$ satisfies the *equality (7.25) (in $H_{s-1}$) and the energy identity*

$$\frac{1}{2}\frac{d}{dt}\left(\|u(t)\|_{s+1}^2 + \|\partial_t u(t)\|_s^2\right)$$
$$= -\nu\|\partial_t u(t)\|_s^2 + (g(t), \partial_t u(t))_s \tag{7.26}$$

*If, in addition, $g \in C(\mathbf{R}, H_{s-1})$ then equation (7.25) is satisfied for each $t \in \mathbf{R}$ (in $H_{s-1}$). The estimates (7.20$_k$) are fulfilled for*

$$\|\vec{u}(t)\|_{X_s} := \left[\|u(t)\|_{s+1}^2 + \|\partial_t u(t)\|_s^2\right]^{1/2} .$$

*The estimate*

$$\|\partial_{tt}^2 u(t)\|_{s-1} \le \|u(t)\|_{s+1} + \nu\|\partial_t u(t)\|_{s-1} + \|g(t)\|_{s-1} \tag{7.27}$$

*is also true.*
The last statement follows from (7.25). The following theorem holds:

### Theorem 7.3

*If the function $g: \mathbf{R} \to H_r$ is absolutely continuous and $\partial_t g \in L_{p,loc}(\mathbf{R}, H_r)$, $r \in \mathbf{R}$, $p \ge 1$, then for the solution $\vec{u} = \binom{u_0}{u_1}$ of problem (7.11) with $\vec{\Psi} = 0$, $\partial_t \vec{u} \in C(\mathbf{R}, X_r)$, $u_0 \in C(\mathbf{R}, H_{r+2})$ and for $t \in \mathbf{R}^+$ the estimates*

$$\|\partial_t \vec{u}(t)\|_{X_r} \le m_3(\alpha, r)e^{-\alpha t}\|g(0)\|_r$$
$$+ m_4(\alpha, r)\int_0^t e^{-\alpha(t-\tau)}\|\partial_t g(\tau)\|_r d\tau , \tag{7.28}$$

$$\|u_0(t)\|_{r+2} \le c\|\partial_t \vec{u}(t)\|_{X_r} + \|g(t)\|_r , \tag{7.29}$$

*hold. Here $c = \sqrt{2}\max\{1, \nu\lambda_1^{-1/2}\}$.*
The proof is based on Theorems 7.1 and 7.2. Theorem 7.2 guarantees the existence of $\vec{u} \in C(\mathbf{R}, X_r)$, the estimates (7.20$_k$) with $\vec{u}(0) = 0$ and $s = r$, as well as $\partial_t \vec{u} \in C(\mathbf{R}, X_{r-1})$.

The latter implies $a\partial_t \vec{u} = \partial_t a\vec{u} \in C(\mathbf{R}, X_{r-2})$ and from $\vec{u} \in C(\mathbf{R}, X_r)$ it follows that $a\vec{u} \in C(\mathbf{R}, X_{r-1})$. So, all the members in the equation (7.11) are elements of $C(\mathbf{R}, X_{r-1})$. Furthermore, since both terms on the right hand side of (7.11) are differentiable in $t$ and their derivatives are elements of $C(\mathbf{R}, X_{r-2})$ and $L_{p,loc}(\mathbf{R}, X_r)$, respectively, then $\partial_t \vec{u}$ also is differentiable in $t$, $\partial_{tt} \vec{u} \in L_{p,loc}(\mathbf{R}, X_{r-2})$ and $\partial_t \vec{u}$ is a solution of the problem

$$\partial_t \vec{w}(t) = a\vec{w}(t) + \partial_t \vec{g}(t), \quad \vec{w}|_{t=0} = \vec{g}(0) \tag{7.30}$$

belonging to $C(\mathbf{R}, X_{r-1})$. On the other hand, the same Theorem 7.2, when applied to the problem (7.30), guarantees the existence of the solution $\vec{w} \in C(\mathbf{R}, X_r)$ as $\vec{g}(0) \in X_r$ and $\partial_t \vec{g} \in L_{p,loc}(\mathbf{R}, H_r)$.

Since $\partial_t \vec{u}$ and $\vec{w}$ are both weak solutions of (7.30), they must coincide according to the uniqueness theorem, i.e. $\partial_t \vec{u} = \vec{w} \in C(\mathbf{R}, X_r)$. The estimate (7.28) is nothing but the estimate $(7.20_1)$ (with $s = r$) for $\vec{w}$ and the estimate (7.29) follows from the fact that $\vec{u}(t) = \binom{u_0(t)}{u_1(t)} = \binom{u(t)}{\partial_t u(t)}$, where $u(t)$ is the solution of (7.25).

### 7.3 On global unique solvability of the nonlinear problem

In this section we investigate the global unique solvability of problem (7.1) or, what is the same, of problem (7.3). We shall use both of these formulations and the relations $v(t) = v_0(t)$, $\partial_t v(t) = v_1(t)$ between the solutions $v(t)$ of (7.1) and $\vec{v}(t) = \binom{v_0(t)}{v_1(t)}$ of (7.3). We choose $X_0$ as the phase-space and (according to this choice) impose on $f$ the following conditions:

(a) $f : H_1 \to H$, $f(0) = 0$ and for all $u_1, u_2 \in H_1$
$$\|f(u_1) - f(u_2)\| \le \Phi_1(\max\{\|u_1\|_1; \|u_2\|_1\})\|u_1 - u_2\|_1. \quad (7.31)$$
Here and below, $\Phi_k$ are some functions, $\Phi_k : \mathbf{R}^m \to \mathbf{R}^+$, their explicit form being insignificant; we only assume that these functions are continuous and nondecreasing when each of their arguments increases.

(b) For any $k = 1, 2, \ldots$ and all $u_1, u_2 \in H_1$
$$\begin{aligned} |(f(u_1) - f(u_2), \phi_k)| \\ \le \Phi_2(k, \max\{\|u_1\|_1, \|u_2\|_1\})\|u_1 - u_2\| \, ; \end{aligned} \quad (7.32)$$

(c) $f$ is a potential operator with the continuous potential $\mathcal{F} : H_1 \to \mathbf{R}$, $\mathcal{F}(0) = 0$ such that for every $u \in C(\mathbf{R}, H_1)$ with $\partial_t u \in C(\mathbf{R}, H)$ the function $\mathcal{F}(u(.)) : \mathbf{R} \to \mathbf{R}$ is absolutely continuous and
$$\frac{d}{dt} \mathcal{F}(u(t)) = (f(u(t)), \partial_t u(t)) . \quad (7.33)$$
Moreover, for all $u \in H_1$, $\mathcal{F}(u)$ must satisfy the inequality
$$-\mathcal{F}(u) \le (\frac{1}{2} - \nu_1)\|u\|_1^2 + c_1 \quad (7.34)$$
with some $\nu_1 \in (0, \frac{1}{2}]$ and $c_1 \in \mathbf{R}^+$.

From (7.33) it follows that
$$\mathcal{F}(u) = \int_0^1 (f(tu), u)dt , \quad \forall u \in H_1 \quad (7.35)$$

and from (7.35) and (7.31) the estimate

$$|\mathcal{F}(u)| \leq \frac{1}{2}\Phi_1(\|u\|_1)\|u\|_1\|u\|$$

$$\leq \frac{1}{2\sqrt{\lambda_1}}\Phi_1(\|u\|_1)\|u\|_1^2 \equiv \Phi_3(\|u\|_1) \tag{7.36}$$

follows.

The following theorem holds:

**Theorem 7.4**

*Let $h \in H$, $\vec{\Psi} \in X_0$ and $f$ satisfy the conditions (a)–(c). Then problem (7.3) has the unique solution $\vec{v}$ enjoying the following properties: $\vec{v} \in C(\mathbf{R}, X_0)$, $\partial_t \vec{v} \in C(\mathbf{R}, X_{-1})$; equation (7.3) is fulfilled for all $t \in \mathbf{R}$ in $X_{-1}$; $\vec{v}(t)$ is given by*

$$\vec{v}(t) = U_t(\vec{v}(0)) + \int_0^t U_{t-\tau}\begin{pmatrix} 0 \\ -f(v_0(\tau)) + h \end{pmatrix} d\tau, \tag{7.37}$$

*where $U_t \in L(X_0 \to X_0)$ are the solution operators of problem (7.22); the energy relation holds:*

$$\frac{\mathrm{d}}{\mathrm{d}t}\mathcal{L}(\vec{v}(t)) = -\nu\|v_1(t)\|^2, \ t \in R, \tag{7.38}$$

*where the function $\mathcal{L}: X_0 \to \mathbf{R}$ is defined by*

$$\mathcal{L}(\vec{u}) := \frac{1}{2}\|u_0\|_1^2 + \frac{1}{2}\|u_1\|^2 + \mathcal{F}(u_0) - (h, u_0),$$

$$\vec{u} = \begin{pmatrix} u_0 \\ u_1 \end{pmatrix}. \tag{7.39}$$

*For the solution $\vec{v}(t)$ the estimates (7.45)–(7.46$_k$) hold.*

*The solution operators $V_t$ of problem (7.3) form a continuous group $\{V_t, t \in \mathbf{R}, X_0\}$; the corresponding semigroup $\{V_t, t \in \mathbf{R}^+, X_0\}$ is bounded.*

First we obtain the basic a priori estimate for the solutions of problem (7.3). To this end we rewrite (7.3) in the form

$$\partial_t \vec{v}(t) = \alpha \vec{v}(t) + \vec{g}(t), \quad \vec{v}|_{t=0} = \vec{\Psi}, \tag{7.40}$$

where

$$\vec{g}(t) \equiv \vec{g}(t; \vec{\Psi}) := \begin{pmatrix} 0 \\ g(t; \Psi) \end{pmatrix}$$

$$= \vec{f}(\vec{v}(t)) + \vec{h} = \begin{pmatrix} 0 \\ -f(v_0(t)) \end{pmatrix} + \begin{pmatrix} 0 \\ h \end{pmatrix}. \tag{7.41}$$

The dependence of $\vec{g}$ on the functional parameter $\vec{\Psi}$ is due to the fact that the solution $\vec{v}$ of problem (7.3) depends on $\vec{\Psi}$.

Equation (7.40) has the form (7.11) and therefore the solutions $\vec{v}$ of (7.40) satisfies (7.12) with $s = 0$ (provided $\vec{v}$ is smooth enough, see above). The last term in (7.12), because of condition (c) with $v_1(t) = \partial_t v_0(t)$, is equal to

$$(g(t), v_1(t)) = (-f(v_0(t)) + h \,, \partial_t v_0(t))$$
$$= +\frac{\mathrm{d}}{\mathrm{d}t}[-\mathcal{F}(v_0(t)) + (h, v_0(t))] \,. \tag{7.42}$$

Therefore, (7.38) is nothing but the relation (7.12) with $s = 0$.

From (7.38) and (7.36) it follows

$$\mathcal{L}(\vec{v}(t)) + \nu \int_0^t \|v_1(\tau)\|^2 \mathrm{d}\tau$$
$$= \mathcal{L}(\vec{v}(0)) \tag{7.43}$$
$$\leq \frac{1}{2}\|\Psi_1\|^2 + \frac{1}{2}\|\Psi_0\|_1^2 + \Phi_3(\|\Psi_0\|_1) + \|h\|\|\Psi_0\| \,.$$

On the other hand, from (7.34) we deduce

$$\mathcal{L}(\vec{v}(t))$$
$$\geq \frac{1}{2}\|v_1(t)\|^2 + \frac{1}{2}\|v_0(t)\|_1^2$$
$$- (\frac{1}{2} - \nu_1)\|v_0(t)\|_1^2 - c_1 - \|h\|\lambda_1^{-1/2}\|v_0(t)\|_1 \tag{7.44}$$
$$\geq \frac{1}{2}\|v_1(t)\|^2 + \frac{\nu_1}{2}\|v_0(t)\|_1^2 - c_1 - (2\nu_1\lambda_1)^{-1}\|h\|^2 \,.$$

Now (7.43) and (7.44) give the "a priori" estimate:

$$\|v_1(t)\|^2 + \nu_1\|v_0(t)\|_1^2 + 2\nu \int_0^t \|v_1(\tau)\|^2 \mathrm{d}\tau$$
$$\leq 2c_1 + (\nu_1\lambda_1)^{-1}\|h\|^2 + \|\Psi_1\|^2 + \|\Psi_0\|_1^2 \tag{7.45}$$
$$+ 2\Phi_3(\|\Psi_0\|_1) + 2\|h\|\|\Psi_0\|_1\lambda_1^{-1/2}$$
$$\equiv \Phi_4(\|\vec{\Psi}\|_{X_0}) \,.$$

In particular,

$$\|\vec{v}(t)\|_{X_0} \leq \Phi_5(\|\vec{\Psi}\|_{X_0}), \quad t \in \mathbf{R}^+ \,, \tag{$7.46_1$}$$

$$\|\vec{v}(t)\|_{X_0} \leq \Phi_6(|t|, \|\vec{\Psi}\|_{X_0}), \quad t \in \mathbf{R}^- \,. \tag{$7.46_2$}$$

The explicit form of the functions $\Phi_5$ and $\Phi_6$ is not significant to our purpose; note that $\Phi_6(|t|, \xi) \to +\infty$ when $t \to -\infty$.

We shall construct the approximate solutions $\vec{v}^m(t)$, $m = 1, 2, \ldots$ of (7.3) by means of the Galerkin-Faedo method using the eigenfunctions $\{\phi_k\}_{k=1}^\infty$ of the operator $A$ as the basis in $H_s$. Namely, we look for

$\vec{v}^m(t) = \binom{v^m(t)}{\partial_t v^m(t)}$ where $v^m(t) = \sum_{\ell=1}^m c_\ell^m(t)\phi_\ell$ and the coefficients $c_\ell^m(t) = (v^m(t), \phi_\ell)$, $\ell = 1, \ldots, m$ are the solutions of the Cauchy problem for the following system of $m$ ordinary differential equations:

$$\partial_{tt}^2(v^m(t), \phi_k) + \nu \partial_t(v^m(t), \phi_k)$$
$$+ \lambda_k(v^m(t), \phi_k) + (f(v^m(t)), \phi_k) = (h, \phi_k) \qquad (7.47)$$
$$c_k^m(0) = (\Psi_0, \phi_k), \ \partial_t c_k^m(0) = (\Psi_1, \phi_k), \ k = 1, \ldots, m.$$

Let us prove that the "a priori" estimates $(7.46_k)$ hold for $\vec{v}^m(t) := \binom{v^m(t)}{\partial_t v^m(t)}$, and, hence, for any $T \in \mathbf{R}^+$:

$$\max_{t \in [-T,T]} \|\vec{v}^m(t)\|_{X_0} \le \Phi_7(T, \|\vec{\Psi}\|_{X_0}), \ m = 1, 2, \ldots, \qquad (7.48)$$

where $\Phi_7(T, \|\vec{\Psi}\|_{X_0}) = \max\{\Phi_5(\|\vec{\Psi}\|_{X_0}); \ \Phi_6(T; \|\vec{\Psi}\|_{X_0})\}$.

We multiply the $k$-th equations (7.47) by $\partial_t c_k^m(t)$ and sum the results for $k = 1, \ldots, m$. It is easy to verify that this yields for $\vec{v}^m(t)$ the equality (7.38), from which the estimates $(7.46_k)$ follow. The estimates (7.48) for the solutions of (7.47) and our hypothesis on $f$ guarantee the global unique solvability of the problem (7.47). The coefficients $c_\ell^m(t)$ are twice continuously differentiable in $t$ and, hence, $v^m(t)$ and $\partial_t v^m(t)$ belong to $C(\mathbf{R}, X_0)$. To perform the limit along some subsequence $m_j \to +\infty$, let us estimate the quantities $|(\partial_{tt}^2 v^m(t), \phi_k)|$, using (7.32) and (7.47). Namely,

$$|(\partial_{tt}^2 v^m(t), \phi_k)|$$
$$\le \nu\|\partial_t v^m(t)\| + \lambda_k^{1/2}\|v^m(t)\|_1$$
$$+ \Phi_2(k, \|v^m(t)\|_1)\|v^m(t)\| + \|h\| \qquad (7.49)$$
$$\equiv \Phi_8(k, \|\vec{v}^m(t)\|_{X_0}).$$

By (7.48) and (7.49) we may choose from $\{\vec{v}^m(t)\}_{m=1}^\infty$ a subsequence $\{\vec{v}^{m_j}(t)\}_{j=1}^\infty$ enjoying the following properties:

(1) $\{v^{m_j}(\cdot)\}$ converges to some element $v \in C(\mathbf{R}, H)$ in the norms of spaces $C([-T,T], H)$, with any $T \in \mathbf{R}^+$;

(2) $\{v^{m_j}(\cdot)\}$ and $\{\partial_t v^{m_j}(\cdot)\}$ converge weakly in the Hilbert spaces $L_2((-T,T), H_1)$ and $L_2((-T,T), H)$, $\forall T \in \mathbf{R}^+$, respectively; hence, $v \in L_{2,loc}(\mathbf{R}, H_1)$ and $\partial_t v \in L_{2,loc}(\mathbf{R}, H)$ and for almost every $t$

$$\|v(t)\|_1 \le \varliminf_{j \to \infty} \|v^{m_j}(t)\|_1 \quad \text{and} \quad \|\partial_t v(t)\| \le \varliminf_{j \to \infty} \|\partial_t v^{m_j}(t)\|.$$

(3) For any (fixed) $k = 1, 2, \ldots, \{\partial_t c_k^{m_j}(\cdot) = (\partial_t v^{m_j}(\cdot), \phi_k)\}_{j=1}^\infty$

converges in $C([-T,T])$, $\forall T \in \mathbf{R}^+$, to $\partial_t c_k(\cdot) \equiv (\partial_t v(\cdot), \phi_k) \in C(R)$.

From (1) and the hypothesis (7.32) it follows that $(f(v^{m_j}(\cdot)), \phi_k)$ converges (as $j \to +\infty$) to $(f(v(\cdot), \phi_k)$ in $C([-T,T])$, $\forall T \in \mathbf{R}^+$. Thus, for a fixed $k$, all terms in (7.47) except the first one have the appropriate limits in $C([-T,T])$, $\forall T \in \mathbf{R}^+$. Hence, $\partial_{tt}^2(v^m(\cdot), \phi_k)$ also has a limit (in the same sense) and this limit is $\partial_{tt}^2(v(\cdot), \phi_k) \in C([-T,T])$, $\forall T \in \mathbf{R}^+$. Now, we may conclude that the limit $v(\cdot)$ satisfies the equations:

$$
\begin{aligned}
&\partial_{tt}^2(v(t), \phi_k) + \nu \partial_t(v(t), \phi_k) \\
&\quad + \lambda_k(v(t), \phi_k) + (f(v(t)), \phi_k) = (h, \phi_k)\,, \\
&(v, \phi_k)|_{t=0} = (\Psi_0, \phi_k), \partial_t(v, \phi_k)|_{t=0} = (\Psi_1, \phi_k)\,, \\
&\hspace{4cm} k = 1, 2, \ldots,
\end{aligned}
\tag{7.50}
$$

and, besides, $v \in C(\mathbf{R}, H) \cap L_{2,loc}(\mathbf{R}, H_1)$ and $\partial_t v \in L_{\infty,loc}(\mathbf{R}, H)$. Moreover, $g(\cdot, \vec{\Psi}) \equiv -f(v(\cdot)) + h$ is an element of $L_{\infty,loc}(\mathbf{R}, H)$ and

$$
\begin{aligned}
\|g(t, \vec{\Psi})\| &\leq \Phi_1(\|v(t)\|_1)\|v(t)\|_1 + \|h\| \\
&\leq \Phi_1(\Phi_7(T, \|\vec{\Psi}_0\|_{X_0}))\Phi_7(T, \|\vec{\Psi}_0\|_{X_0}) + \|h\| \quad (7.51) \\
&\equiv \Phi_8(T, \|\vec{\Psi}_0\|_{X_0})\,,
\end{aligned}
$$

for $t \in [-T, T]$. Hence $\vec{v}(\cdot) \equiv \left(\begin{smallmatrix} v(\cdot) \\ \partial_t v(\cdot) \end{smallmatrix}\right)$ may be interpreted as a weak solution of the linear problem (7.40) with $g \in L_{\infty,loc}(\mathbf{R}, H)$ and $\vec{\Psi} \in X_0$. On the other hand Theorem 7.2 guarantees for this problem the existence of a solution in $C(\mathbf{R}, X_0)$, and because of Theorem 7.1 this solution must coincide with $\vec{v}$. Using this result and, in addition the hypotheses on $f$ and applying Theorem 7.2 we obtain all the statements of Theorem 7.4 except the last one and the uniqueness for problem (7.3). In order to prove the uniqueness suppose that the problem (7.3) has another solution $\vec{v}'(t)$ possessing all the properties of $\vec{v}(t)$. The difference $\vec{w}(t) = \vec{v}(t) - \vec{v}'(t)$ may be regarded as a solution of the linear problem (7.40) with the free term $\vec{g}(t) = \left(\begin{smallmatrix} 0 \\ -f(v_0(t))+f(v_0'(t)) \end{smallmatrix}\right)$ and $\vec{\Psi} \equiv 0$. Since $\vec{w} \in C(\mathbf{R}, X_0)$ and $\vec{g}(\cdot) \in L_{\infty,loc}(\mathbf{R}, X_0)$, for $\vec{w}$ the energy equality (7.12) holds. We estimate the last term in (7.12) using (7.31) and (7.46$_k$):

$$
\begin{aligned}
|(-f(v_0(t)) &- f(v_0'(t)), w_1(t))| \\
&\leq \Phi_1(\max\{\|v_0(t)\|_1, \|v_0'(t)\|_1\})\|w_0(t)\|_1\|w_1(t)\| \quad (7.52) \\
&\leq \Phi_9(|t|)\|\vec{w}(t)\|_{X_0}^2\,.
\end{aligned}
$$

Hence $\vec{w}(t)$ satisfies the inequality

$$\frac{1}{2}\frac{d}{dt}\|\vec{w}(t)\|^2_{X_0} + \nu\|w_1(t)\|^2 \le \Phi_9(|t|)\|\vec{w}(t)\|^2_{X_0} \qquad (7.53)$$

and since $\vec{w}(0) = 0$ we conclude that $\vec{w}(t) \equiv 0$. Consequently, problem (7.3) has the unique solution with the properties indicated in Theorem 7.4.

Thus we have proved that the solution operators $V_t : \vec{\Psi} \in X_0 \to \vec{v}(t) \in X_0$ are defined for all $\vec{\Psi} \in X_0$ and $t \in \mathbf{R}$; they are single-valued and because of (7.45), (7.46$_k$) are bounded (i.e. they map bounded sets into bounded sets). The family $\{V_t, t \in \mathbf{R}, X_0\}$ has the group property: $V_{t_1+t_2} = V_{t_1} V_{t_2}$ for all $t_1, t_2 \in \mathbf{R}$. In order to prove that the operators $V_t$ are continuous on $X_0$, let $\vec{v}(t) = V_t(\vec{\Psi})$ and $\vec{\tilde{v}}(t) = V_t(\vec{\tilde{\Psi}})$. As in the proof of uniqueness, we have for $\vec{w}(t) = \vec{v}(t) - \vec{\tilde{v}}(t)$ the inequality (7.53) with

$$\Phi_9(|t|) = \frac{1}{2}\Phi_1(\max\{\|v_0(t)\|_1, \|\tilde{v}_0(t)\|_1\}) \, .$$

By (7.46$_k$), $\Phi_9(|t|) \le \Phi_{10}(|t|, \rho)$, where $\rho \equiv \max\{\|\vec{\Psi}\|_{X_0}, \|\vec{\tilde{\Psi}}\|_{X_0}\}$ and consequently

$$\frac{1}{2}\frac{d}{dt}\|\vec{w}(t)\|^2_{X_0} + \nu\|w_1(t)\|^2 \le \Phi_{10}(|t|, \rho)\|\vec{w}(t)\|^2_{X_0} \, . \qquad (7.54)$$

The integration of this inequality yields

$$\|\vec{w}(t)\|_{X_0} \equiv \|V_t(\vec{\Psi}) - V_t(\vec{\tilde{\Psi}})\|_{X_0} \le \Phi_{11}(|t|, \rho)\|\vec{\Psi} - \vec{\tilde{\Psi}}\|_{X_0} (7.55)$$

for all $\vec{\Psi}, \vec{\tilde{\Psi}}$ belonging to the ball $B_\rho(0) \subset X_0$ of radius $\rho$. This inequality guarantees the uniform continuity of $V_t$ on every ball $B_\rho(0)$. Since $V_t(\vec{\Psi})$ is continuous in $t$ for each $\vec{\Psi} \in X_0$, $V_t(\vec{\Psi})$ is continuous in $(t, \vec{\Psi}) \in \mathbf{R} \times X_0$, i.e. the group $\{V_t, t \in \mathbf{R}, X_0\}$ is continuous.

Finally the boundedness of the semigroup $\{V_t, t \in \mathbf{R}^+, X_0\}$ follows from the estimate (7.46$_1$).

## 7.4 On differentiability of solution operators

Let us consider the differentiability of the solution operators $V_t$ of problem (7.3). Assume that the conditions of Theorem 3.4 hold and, in addition, $f$ is differentiable in the following sense:

(d) for all $u, \hat{u} \in H_1$

$$f(\hat{u}) - f(u) = f'(u)(\hat{u} - u) + r_f(\hat{u}, u) \qquad (7.56)$$

where $f'(u) \in L(H_1 \to H)$ and

$$\|f'(u)\|_{L(H_1 \to H)} \leq \Phi_{12}(\|u\|_1) ; \tag{7.57}$$

the remainder $r_f(\hat{u}, u)$ satisfies the inequality

$$\|r_f(\hat{u}, u)\| \leq \Phi_{13}(\max\{\|\hat{u}\|_1, \|u\|_1\})\gamma(\|\hat{u} - u\|_1)\|\hat{u} - u\|_1 \tag{7.58}$$

where $\gamma : \mathbf{R}^+ \to \mathbf{R}^+$ is a continuous function, $\gamma(0) = 0$.

Due to hypothesis (a) (see (7.31)) condition (d) is a new restriction only for $\hat{u}$ close to $u$.

In this situation we shall prove that $V_t$ is differentiable in the following sense:

$$V_t(\vec{\Psi} + s\vec{\xi}) - V_t(\vec{\Psi}) = sU(t, \vec{\Psi})\vec{\xi} + \vec{R}(t, s, \vec{\Psi}, \vec{\xi}) \tag{7.59}$$

for $\vec{\Psi} \in X_0$, $s \in [-s_0, s_0]$ and $\vec{\xi} \in X_0$ with $\|\vec{\xi}\|_{X_0} = 1$. Here $U(t, \vec{\Psi}) \in L(X_0 \to X_0)$ and $\mathbf{R}(\ldots)$ satisfy the inequality

$$\|\vec{R}(t, s, \vec{\Psi}, \vec{\xi})\|_{X_0} \leq \Phi_{14}(|t|, |s|, \|\vec{\Psi}\|_{X_0}) , \tag{7.60}$$

where $\Phi_{14} : \mathbf{R}^3 \to \mathbf{R}^+$ is a continuous nondecreasing function of its arguments and $|s|^{-1}\Phi_{14}(|t|, |s|, \|\vec{\Psi}\|_{X_0}) \to 0$ when $s \to 0$. The operator $U(t, \vec{\Psi}) : X_0 \to X_0$ is the differential of $V_t$ at the point $\vec{\Psi}$. We shall use the notation $\vec{v}(t, \vec{\Psi})$ for $V_t(\vec{\Psi})$, and $v_k(t, \vec{\Psi})$, $k = 0, 1$, for its components.

If the representation (7.59) really holds then it is easy to verify that $U(t, \vec{\Psi})\vec{\xi}$ is a solution of the linear problem

$$\partial_t \vec{u}(t) = a\vec{u}(t) + B(t, \vec{\Psi})\vec{u}(t) , \quad \vec{u}|_{t=0} = \vec{\xi} , \tag{7.61}$$

where $B(t, \vec{\Psi})\vec{u}(t) = \begin{pmatrix} 0 \\ -f'(v(t, \vec{\Psi}))u_0(t) \end{pmatrix}$.

Formally, to obtain (7.61) one has to subtract from the equation (7.3) for $V_t(\vec{\Psi} + s\vec{\xi})$ the same equation written for $V_t(\vec{\Psi})$, to divide the result by $s$ and then pass to the limit when $s \to 0$. Equation (7.61) is the so-called equation in variations for (7.3) (on the solution $v(t, \vec{\Psi})$).

It is easier to obtain the equation in variations for the scalar problem (7.1). This equation is

$$\partial_{tt}^2 u(t) + \nu \partial_t u(t) + Au + f'(v_0(t, \Psi))u(t) = 0 \tag{7.61'}$$

The relations between $u(t)$ and $\vec{u}(t)$ are $u(t) = u_0(t)$, $\partial_t u(t) = u_1(t)$. Equation (7.61) is the vector form of equation (7.61').

The hypothesis (7.57) implies that $B(t, \vec{\Psi})\vec{u} \in X_0$ for any $\vec{u} \in X_0$ and $\|f'(v_0(t, \vec{\Psi}))u_0\| \leq \Phi_{12}(\|v_0(t, \vec{\Psi})\|_1)\|u_0\|_1$.

The unique solvability of problem (7.61) for any $\vec{\xi} \in X_0$ and $\vec{\Psi} \in X_0$ is proved as for problem (7.3) (see Theorem 7.4) but with essential simplifications due to the linearity of problem (7.61). Its solution $\vec{u}(t)$ enjoys the following properties:

$$\vec{u} \in C(\mathbf{R}, X_0), \qquad B(t, \vec{\Psi})\vec{u}(t) \in X_0,$$
$$\|B(t, \vec{\Psi})\vec{u}(t)\|_{X_0} \le \Phi_{15}(|t|, \|\vec{\Psi}\|_{X_0}),$$
$$\partial_t \vec{u} \in L_{\infty, loc}(\mathbf{R}, X_{-1});$$

equation (7.61) is fulfilled in $X_{-1}$ for all $t \in \mathbf{R}$ and the main energy relation

$$\frac{1}{2}\frac{d}{dt}\|\vec{u}(t)\|^2_{X_0} = -\nu\|u_1(t)\|^2 - (f'(v_0(t, \vec{\Psi}))u_0(t), u_1(t)) \quad (7.62)$$

holds. From (7.62) and from the assumption (7.57) the estimate

$$\|U(t, \vec{\Psi})\|_{L(X_0 \to X_0)} \le \Phi_{16}(|t|, \|\vec{\Psi}\|_{X_0}) \qquad (7.63)$$

is derived in the standard way.

Generally, $\Phi_{16} \to +\infty$ as $|t| \to +\infty$.

Thus we may take this fact for granted and estimate the remainder term

$$\vec{R}(t, s, \vec{\Psi}, \vec{\xi}) \equiv \vec{v}(t, \vec{\Psi} + s\vec{\xi}) - \vec{v}(t, \vec{\psi}) - s\vec{u}(t, \vec{\Psi}, \vec{\xi}),$$

where $\vec{u}(t, \vec{\Psi}, \vec{\xi}) \equiv U(t, \vec{\Psi})\vec{\xi}$. It is clear that $\vec{R} \in C(\mathbf{R}, X_0)$. If we subtract from equation (7.3) for $\vec{v}(t, \vec{\Psi} + s\vec{\xi})$ the same equation for $\vec{v}(t, \vec{\Psi})$ and $s$ times equation (7.61) for $\vec{u}(t, \vec{\Psi}, \vec{\xi})$, we obtain for $\vec{R}(\ldots)$ :

$$\partial_t \vec{R}(t, \ldots) = a\vec{R}(t, \ldots) + \vec{B}_1(t), \qquad (7.64)$$

where $\vec{B}_1(t) = \begin{pmatrix} 0 \\ B_1(t, \ldots) \end{pmatrix}$ and

$$B_1(t, \ldots) \equiv -f(v_0(t, \vec{\Psi} + s\vec{\xi})) + f(v_0(t, \vec{\Psi}))$$
$$+ sf'(v_0(t, \vec{\Psi}))u_0(t, \vec{\Psi}, \vec{\xi}).$$

Moreover, it is evident that

$$\vec{R}|_{t=0} = 0. \qquad (7.65)$$

From the estimates (7.46$_1$) and (7.63) and the assumption (7.57) we deduce that $B_1 \in L_{\infty, loc}(\mathbf{R}, H)$. The vector-valued function $\vec{R}: t \to X_0$ may be regarded as a solution (from $C(\mathbf{R}, X_0)$) of the linear problem (7.11) with the free term $\vec{g}(t) = \vec{B}_1(t)$ and with zero initial value.

According to Theorem 7.2 the following energy relation holds:

$$\frac{1}{2}\frac{d}{dt}\|\vec{R}(t, \ldots)\|^2_{X_0}$$
$$= -\nu\|R_1(t, \ldots)\|^2 + (B_1(t, \ldots), R_1(t, \ldots)). \qquad (7.66)$$

$B_1(t,\ldots)$ using (7.56) may be represented in the form

$$B_1(t,\ldots) = -f'(v_0(t,\vec{\Psi}))R_0(t,s,\vec{\Psi},\vec{\xi})$$
$$- r_f(v_0(t,\vec{\Psi}+s\vec{\xi}), v_0(t,\vec{\Psi})),$$

where $R_0(t,\ldots)$ is the first component of the vector $\vec{R}(t,\ldots)$; note that the second component $R_1(t,\ldots)$ appeared in (7.66). The last term in (7.66) may be estimated as follows:

$$|(B_1(t,\ldots), R_1(t,\ldots))|$$
$$\leq \Phi_{12}(\|v_0(t,\vec{\Psi})\|_1)\|R_0(t,\ldots)\|_1\|R_1(t,\ldots)\|$$
$$+ \Phi_{13}(\max\{\|v_0(t,\vec{\Psi}+s\vec{\xi})\|_1, \|v_0(t,\vec{\Psi})\|_1\})$$
$$\cdot \gamma(\|v_0(t,\vec{\Psi}+s\vec{\xi}) - v_0(t,\vec{\Psi})\|_1) \qquad (7.67)$$
$$\cdot \|v_0(t,\vec{\Psi}+s\vec{\xi}) - v_0(t,\vec{\Psi})\|_1\|R_1(t,\ldots)\|$$
$$\leq \Phi_{17}(|t|, \|\vec{\Psi}\|_{X_0})\|\vec{R}(t,\ldots)\|^2_{X_0}$$
$$+ \Phi_{18}(|t|, |s|, \|\vec{\Psi}\|_{X_0})\|\vec{R}(t,\ldots)\|_{X_0},$$

where $\Phi_{18}(\ldots)$ is such that $|s|^{-1}\Phi_{18}(|t|, |s|, \|\vec{\Psi}\|_{X_0}) \to 0$ when $s \to 0$. We have used the assumptions (7.57), (7.58), the estimates $(7.46_k)$ for the solutions $\vec{v}(t,\vec{\Psi}+s\vec{\xi})$ and $\vec{v}(t,\vec{\Psi})$, and the estimate (7.55) for their difference. Substituting (7.67) into (7.66) and integrating the resulting inequality, keeping in mind (7.65), we obtain the desired estimate (7.60). Let us summarize the obtained results in the following theorem.

### Theorem 7.5

*Let the assumptions of Theorem 3.4 and the condition (d) be fulfilled. Then for all $t \in \mathbf{R}$ the operator $V_t$ is differentiable at each point $\vec{\Psi} \in X_0$, its differential $U(t,\vec{\Psi})$ belongs to the class $L(X_0 \to X_0)$ and the norms $\|U(t,\vec{\Psi})\|_{L(X_0 \to X_0)}$ are uniformly bounded for $\vec{\Psi} \in B_\rho(0) \subset X_0, \forall\rho > 0$. The estimates (7.60) hold.*

### 7.5 On the affiliation of the semigroup $\{V_t, t \in \mathbf{R}^+, X_0\}$ to the class $A\mathcal{K}$ and attractors

We shall prove the following

### Theorem 7.6

*Let the conditions of Theorem 7.4 and the following condition be satisfied*

*(e) At every point of $u \in H_1$ $f$ has a derivative $f'(u) \in$ $L(H \to H_{-1+\delta})$ where $\delta$ is a number belonging to $(0,1]$ and*

$$\|f'(u)\|_{L(H \to H_{-1+\delta})} \le \Phi_{19}(\|u\|_1) . \qquad (7.68)$$

*Then the semigroup $\{V_t, t \in \mathbf{R}^+, X_0\}$ belongs to the class $A\mathcal{K}$.*
Let $\vec{v}(t, \vec{\Psi})$ be a solution of problem (7.3) guaranteed by Theorem 7.4. Its representation (7.37) holds and it may be interpreted as the representation of the operators $V_t: X_0 \to X_0$ in the form

$$V_t = U_t + W_t, \quad t \in \mathbf{R}^+ , \qquad (7.69)$$

where $U_t: X_0 \to X_0$ are the resolving operators of the linear problem (7.22) and $W_t: X_0 \to X_0$ are defined by the equality

$$W_t(\vec{\Psi}) = \int_0^t U_{t-\tau} \begin{pmatrix} 0 \\ -f(v_0(\tau, \vec{\Psi})) + h \end{pmatrix} d\tau . \qquad (7.70)$$

Let us show that (7.69) allows us to use Theorem 3.3, and thus to prove that the semigroup $\{V_t, t \in \mathbf{R}^+, X_0\}$ belongs to the class $A\mathcal{K}$.

We already know that $\|U_t\|_{L(X_0 \to X_0)} \to 0$, when $t \to +\infty$; therefore we only have to prove the compactness of the operators $W_t$, $t > 0$. To this end it is sufficient to prove that each $W_t$ ($t > 0$) transforms any set $B$, bounded in $X_0$, into a set bounded in $X_\delta$, $\delta > 0$. We also know that

$$\vec{v}(., \vec{\Psi}) \in C(R^+, X_0) \quad \text{and}$$

$$\sup_{t \in \mathbf{R}^+} \|\vec{v}(t, \vec{\Psi})\|_{X_0} \le \Phi_{20}(\|\vec{\Psi}\|_{X_0}) ,$$

where $\Phi_{20}(\|\vec{\Psi}\|_{X_0}) = (\max\{1; \nu_1^{-1}\} \Phi_4(\|\vec{\Psi}\|_{X_0}))^{1/2}$ (see (7.45)).
From this and from condition (a) it follows that

$$g(\cdot, \vec{\Psi}) \equiv -f(v_0(\cdot, \vec{\Psi})) + h$$

belongs to $C(\mathbf{R}^+, H)$ and

$$\sup_{t \in \mathbf{R}^+} \|g(t, \vec{\Psi})\| \le \sup_{t \in \mathbf{R}^+} \left[ \Phi_1(\|v_0(t, \vec{\Psi})\|_1) \|v_0(t, \vec{\Psi})\|_1 + \|h\| \right] .$$

Additionally, because of condition (e), $g(t, \vec{\Psi})$ is differentiable in $t$,

$$\partial_t g(t, \vec{\Psi}) = -f'(v_0(t, \vec{\Psi})) \partial_t v_0(t, \vec{\Psi})$$
$$= -f'(v_0(t, \vec{\Psi})) v_1(t, \vec{\Psi}) \in H_{-1+\delta}$$

and

$$\|\partial_t g(t, \vec{\Psi})\|_{-1+\delta} \le \Phi_{19}(\|v_0(t, \vec{\Psi})\|_1) \|v_1(t, \vec{\Psi})\| .$$

Therefore $\sup_{t \in \mathbf{R}^+} \|\partial_t g(t, \vec{\Psi})\|_{-1+\delta} \le \Phi_{21}(\|\vec{\Psi}\|_{X_0})$.
We can consider $\vec{W}_t(\vec{\Psi}) \equiv \vec{w}(t, \vec{\Psi})$ as a solution of the linear problem (7.11) with the free term $g(t, \vec{\Psi})$ and zero initial data and use

Theorem 7.3. All its conditions are fulfilled when $r = -1 + \delta$, hence

$$w_0(\cdot, \vec{\Psi}) \in C(R, H_{1+\delta}), w_1(\cdot, \vec{\Psi}) = \partial_t w_0(\cdot, \vec{\Psi}) \in C(\mathbf{R}, H_\delta)$$

and, as is easy to verify, from (7.28) and (7.29) we get the estimate

$$\sup_{t \in R^+} \|W_t(\vec{\Psi})\|_{X_\delta} \leq \Phi_{22}(\|\vec{\Psi}\|_{X_0}), \quad \delta > 0 . \tag{7.71}$$

Thus Theorem 7.6 is proved as $X_\delta$ with $\delta > 0$ is compactly embedded in $X_0$.

Now we shall use the theorems of Part I on attractors for semigroups of the class $A\mathcal{K}$; see Chapter 3. To this end we must prove at least the point-dissipativeness of our semigroup.

Since, in this case, we have the good Lyapunov function $\mathcal{L}$ (see (7.39)) it is sufficient to prove the boundedness of the set $Z$ of all possible stationary points $\vec{z} = \binom{z_0}{z_1}$. Here $z_1 = 0$ and $z_0$ is a solution of problem

$$Az_0 + f(z_0) = h , \quad z_0 \in H_1 . \tag{7.72}$$

Let us assume that $f(\cdot)$ has the property:

(f) for any $u \in H_1$

$$-(f(u), u) \leq (1 - \nu_2)\|u\|_1^2 + c_2, \nu_2 \in (0, 1], c_2 \in \mathbf{R}^+ . \tag{7.73}$$

The inner product in $H$ of (7.72) with $z_0$ and (7.73) give

$$\|z_0\|_1^2 \leq (1 - \nu_2)\|z_0\|_1^2 + c_2 + \|h\|\|z_0\| ;$$

hence the estimate

$$\|z_0\|_1 \leq c_3 \tag{7.74}$$

holds for any possible solution of problem (7.72).

This estimate and the hypothesis (a) on $f$ allow us also to majorize the norm $\|z_0\|_2$, namely

$$\begin{aligned} \|z_0\|_2 = \|Az_0\| &\leq \|f(z_0)\| + \|h\| \\ &\leq \Phi_1(\|z_0\|_1)\|z_0\|_1 + \|h\| \\ &\leq \Phi_1(c_3)c_3 + \|h\| \equiv c_4 . \end{aligned} \tag{7.75}$$

Thus the set $Z$ lies in the ball $B_{c_3}(0) \subset X_0$ and is bounded in $X_1$.

Moreover, $Z$ is closed in $X_0$ (and in $X_1$ as well, see 7.31) and therefore it is compact in $X_0$.

So, if the conditions of Theorem 7.6 and hypothesis (7.73) are fulfilled, then, according to Theorem 7.2, $\{V_t, t \in \mathbf{R}^+, X_0\}$ has the set $Z$ as its attractor $\widehat{\mathcal{M}}$ and the attractor $\mathcal{M}$ is a connected compact set, composed of $Z$ and of all complete trajectories connecting the points of $Z$. Moreover, the attractor $\mathcal{M}$ is a closed bounded subset of the space $X_\delta, \delta > 0$. Indeed, if $\vec{\Psi} \in \mathcal{M}$, then $\vec{v}(t, \vec{\Psi}) \equiv V_t(\vec{\Psi}) \in \mathcal{M}$, for every $t \in \mathbf{R}$, and for arbitrary $t$ and $s$ we have

$$\vec{v}(t, \vec{\Psi}) = U_{t-s}(\vec{v}(s, \vec{\Psi}))$$
$$+ \int_s^t U_{t-\tau} \left( \begin{array}{c} 0 \\ -f(v_0(\tau, \vec{\Psi})) + h \end{array} \right) d\tau . \qquad (7.76)$$

If $s$ tends to $-\infty$, then we get

$$\vec{v}(t, \vec{\Psi}) = \int_{-\infty}^t U_{t-\tau} \left( \begin{array}{c} 0 \\ -f(v_0(\tau, \vec{\Psi})) + h \end{array} \right) d\tau, t \in R , \qquad (7.77)$$

since $\mathcal{M}$ is a bounded set and the operators $U_t$ enjoy the property $(7.23_1)$.

This is the integral equation for complete trajectories lying in $\mathcal{M}$. For $\mathcal{M}$ the following estimate holds

$$\|\mathcal{M}\|_{X_\delta} \leq \Phi_{22}(\|\mathcal{M}\|_{X_0}) , \qquad (7.78)$$

where $\|\mathcal{M}\|_X := \sup_{\vec{\Psi} \in \mathcal{M}} \|\vec{\Psi}\|_X$.

Indeed, let $\vec{\Psi} \in \mathcal{M}$. We can use (7.69) for $V_s$ with $s > 0$ and represent $\vec{\psi}$ as follows:

$$\vec{\Psi} = V_s(\vec{v}(-s, \vec{\Psi})) = U_s(\vec{v}(-s, \vec{\Psi})) + W_s(\vec{v}(-s, \vec{\Psi})) ,$$

where $\vec{v}(-s, \vec{\Psi}) = V_{-s}(\vec{\Psi}) \in \mathcal{M}$.

Then $\sup_{s \in \mathbf{R}^+} \|\vec{v}(-s, \vec{\Psi})\|_{X_\delta} \leq \|\mathcal{M}\|_{X_\delta}$, because $\vec{v}(-s, \vec{\Psi}) \in \mathcal{M}$, and by $(3.23_1)$ $\|U_s(\vec{v}(-s, \vec{\Psi}))\|_{X_\delta} \leq m_4(\alpha, \delta)e^{-\alpha s}\|\mathcal{M}\|_{X_\delta}$. For the second term $W_s(\vec{v}(-s, \vec{\Psi}))$ we have the estimate (7.71) and therefore

$$\|\vec{\Psi}\|_{X_\delta} \leq m_4(\alpha, \delta)e^{-\alpha s}\|\mathcal{M}\|_{X_\delta} + \Phi_{22}(\|\mathcal{M}\|_{X_0}) .$$

If $s$ tends to $+\infty$ then we get (7.78).

So we have proved the following theorem:

### Theorem 7.7

*If $h \in H$ and $f$ satisfies the conditions (a)–(c), (e) and (f) (see (7.73)), then the semigroup $\{V_t, t \in R^+, X_0\}$ has the compact minimal global attractor $\widehat{\mathcal{M}}$ coinciding with the set $Z$ of all stationary points. It is a bounded subset in the space $X_1$. Its minimal global B-attractor $\mathcal{M}$ is connected and compact in $X_0$, consisting of all complete trajectories connecting points of the set $Z$. It is a bounded subset in the space $X_\delta$, $\delta > 0$. For the trajectories $\vec{v}(t, \vec{\Psi}) = V_t(\vec{\Psi})$ on $\mathcal{M}$ the integral equation (7.77) holds.*

Although Theorem 7.7 is meaningful, its practical application for determining $\mathcal{M}$ requires a lot of additional work.

That is why those cases where we can really find a bounded set $B_0$ containing the attractor $\mathcal{M}$ are very important. Now we do this by supposing that $\mathcal{F}$, besides the properties listed above, also enjoys the following:

(g) for all $u \in H_1$

$$\nu_3 \mathcal{F}(u) - (f(u), u) \le c_3 , \tag{7.79}$$

with some $\nu_3 \in (0, 2)$ and $c_3 \in \mathbf{R}^+$.

As in the proof of the estimates $(7.20_k)$ we shall use $\vec{v}(t, \alpha)$ related to the solution $v(t)$ by means of (7.13); $\vec{v}(t, \alpha)$ is the solution of (7.4) and hence the corresponding energy equality of the form (7.16) with $s = 0$ holds. Namely,

$$\begin{aligned}
\frac{1}{2}\frac{\mathrm{d}}{\mathrm{d}t}&\|\vec{v}(t, \alpha)\|_{X_0}^2 \\
&= -\alpha\|\vec{v}_0(t, \alpha)\|_{1,\alpha}^2 - (\nu - \alpha)\|v_1(t, \alpha)\|^2 \\
&\quad + (-f(v_0(t, \alpha)) + h,\ v_1(t, \alpha)) .
\end{aligned} \tag{7.80}$$

The last term may be transformed as follows:

$$\begin{aligned}
(-f(v_0(t, \alpha)) &+ h,\ \alpha v_0(t, \alpha) + \partial_t v_0(t, \alpha)) \\
&= \frac{\mathrm{d}}{\mathrm{d}t}[-\mathcal{F}(v_0(t, \alpha)) + (h, v_0(t, \alpha))] \\
&\quad - \alpha(f(v_0(t, \alpha)),\ v_0(t, \alpha)) + \alpha(h, v_0(t, \alpha)) .
\end{aligned}$$

This enables us to rewrite (7.80) in terms of the function

$$\mathcal{L}_\alpha(\vec{u}) \equiv \frac{1}{2}\|u_0\|_{1,\alpha}^2 + \frac{1}{2}\|u_1\|^2 + \mathcal{F}(u_0) - (h, u_0) .$$

Namely

$$\begin{aligned}
\frac{\mathrm{d}}{\mathrm{d}t}&\mathcal{L}_\alpha(\vec{v}(t, \alpha)) \\
&= -\alpha\|v_0(t, \alpha)\|_{1,\alpha}^2 - (\nu - \alpha)\|v_1(t, \alpha)\|^2 \\
&\quad - \alpha(f(v_0(t, \alpha)), v_0(t, \alpha)) + \alpha(h, v_0(t, \alpha)) \\
&= -\alpha\nu_3 \mathcal{L}_\alpha(\vec{v}(t, \alpha)) \\
&\quad + \alpha[\nu_3 \mathcal{F}(v_0(t, \alpha)) - (f(v_0(t, \alpha)), v_0(t, \alpha))] \\
&\quad + \alpha(1 - \nu_3)(h, v_0(t, \alpha)) \\
&\quad - \alpha(1 - \tfrac{1}{2}\nu_2)\|v_0(t, \alpha)\|_{1,\alpha}^2 \\
&\quad - (\nu - \alpha - \tfrac{1}{2}\alpha\nu_3)\|v_1(t, \alpha)\|^2 .
\end{aligned} \tag{7.81}$$

By the assumption (7.79) the term enclosed in the square brackets [...] on the right hand side of (7.81) is no larger than $\alpha c_3$. The sum of the remaining terms (except the first one) is no larger than $\alpha c_4$, where

$$c_4 = [(2 - \nu_3)\lambda_1(\alpha)]^{-1}(1 - \nu_3)^2\|h\|\,,$$

provided $\alpha$ satisfies the inequality (7.6). Now it follows from (7.81) that

$$\frac{d}{dt}\mathcal{L}_\alpha(\vec{v}(t,\alpha)) \leq -\alpha\nu_3\mathcal{L}_\alpha(\vec{v}(t,\alpha)) + \alpha c_5\,, \tag{7.82}$$

with $c_5 = c_3 + c_4$. By integrating this inequality we obtain

$$\mathcal{L}_\alpha(\vec{v}(t,\alpha)) \leq e^{-\alpha\nu_3 t}\mathcal{L}_\alpha(\vec{v}(0,\alpha)) + \nu_3^{-1}c_5, t \in \mathbf{R}^+\,. \tag{7.83}$$

From (7.83) we want to deduce information about the solution $\vec{v}(t) = C^{-1}(\alpha)\vec{v}(t,\alpha)$ in terms of the norm in $X_0$. We proceed as above, where the estimates $(7.20_k)$ were deduced from $(7.18_k)$. Namely, for arbitrary $\vec{v}, \vec{v}(\alpha) \in X_0$ such that $\vec{v}(\alpha) = C(\alpha)\vec{v}$ we have $(7.19_2)$ with $s = 0$, i.e.

$$m_1^{-1}(0,\alpha)\|\vec{v}\|_{X_0} \leq \|\vec{v}(\alpha)\|_{X_{0,\alpha}} \leq m_2(0,\alpha)\|\vec{v}\|_{X_0}\,. \tag{7.84}$$

Using (7.84) and (7.36) we get

$$\mathcal{L}_\alpha(\vec{v}(\alpha))$$
$$\leq \frac{1}{2}m_2^2(0,\alpha)\|\vec{v}\|_{X_0}^2 + \Phi_3(\|\vec{v}\|_{X_0}) + \|h\|\lambda_1^{-1/2}\|\vec{v}\|_{X_0} \tag{7.85}$$
$$= \Phi_{23}(\|\vec{v}\|_{X_0})\,.$$

In order to estimate $\mathcal{L}_\alpha(\vec{v}(\alpha))$ from below we use (7.84) and (7.34):

$$\mathcal{L}_\alpha(\vec{v}(\alpha)) \geq \frac{1}{2}m_1^{-2}(0,\alpha)\|\vec{v}\|_{X_0}^2 - (\frac{1}{2} - \nu_1)\|\vec{v}\|_{X_0}^2$$
$$- c_1 - \|h\|\lambda_1^{-1/2}\|\vec{v}\|_{X_0}\,. \tag{7.86}$$

In addition to (7.6) we impose on $\alpha$ the following restriction

$$m_1^{-2}(0,\alpha) - 1 + 2\nu_1 = (\mu_1(\alpha)m^{-1}(\alpha))^2 - 1 + 2\nu_1$$
$$\geq 4\nu_4 > 0 \tag{7.87}$$

with some positive $\nu_4$. Since $\mu_1(\alpha)$ and $m(\alpha)$ tend to 1 as $\alpha \to 0$, (7.87) always holds for $\alpha$ small enough. Now, (7.86) and (7.87) yield

$$\mathcal{L}_\alpha(\vec{v}(\alpha)) \geq \nu_4\|\vec{v}\|_{X_0}^2 - c_1 - (4\nu_4\lambda_1)^{-1}\|h\|^2$$
$$\equiv \nu_4\|\vec{v}\|_{X_0}^2 - c_5'\,. \tag{7.88}$$

From the inequalities (7.83), (7.85) and (7.88) it follows that

$$\|\vec{v}(t,\vec{\Psi})\|_{X_0}^2 \leq e^{-\alpha\nu_3 t}\nu_4^{-1}\Phi_{23}(\|\vec{\Psi}\|_{X_0}) + c_6^2, t \in \mathbf{R}^+\,, \tag{7.89}$$

where $c_6^2 = \nu_4^{-1}(c_5' + \nu_3^{-1}c_5)$. This estimate, being true for a solution $\vec{v}(t,\vec{\Psi})$ of (7.3) with arbitrary $\vec{\Psi} \in X_0$, shows that every bounded set

$B \subset X_0$ falls inside the ball

$$B_{2c_6}(0) \equiv \{\vec{v} : \vec{v} \in X_0, \|\vec{v}\|_{X_0} \le 2c_6\} \tag{7.90}$$

after a finite time $t(B)$. Thus we have proved:

### Theorem 7.8

*If $h \in H$ and $f$ satisfies the conditions (a)–(c) and (g), then the semigroup $\{V_t, t \in \mathbf{R}^+, X_0\}$ has the ball $B_{2c_6}(0)$ for a $B$-absorbing set.*

*Remark* A representative class of nonlinear partial differential equations is that in which $f(v)$ is a polynomial $P_m(v)$ of degree $m$. If $m$ is odd and the coefficient of the principal term in $P_m(v)$ is positive, the conditions (7.34) and (7.79) hold with $\nu_1 = \frac{1}{2}$ in (7.34) and $\nu_4 = \frac{1}{m+1}$ in (7.79).

### 7.6 On the dimensions of compact invariant sets

We want here to estimate above $\dim_H(\mathcal{A})$ and $\dim_f(\mathcal{A})$ for a compact invariant set $\mathcal{A} \in X_0$ using Theorems 4.8 and 4.9. Let $f$ satisfy the conditions of Theorem 7.5, its derivative $f'(v_0)$ at each point $v_0 \in \mathcal{P}\mathcal{A} \subset H_1$, where $\mathcal{P}$ is the orthoprojector of $X_0$ onto $H_1$, belongs to $L(H_{1-\beta} \to H_\beta)$ with some $\beta > 0$ and

$$\sup_{v_0 \in \mathcal{P}\mathcal{A}} \|f'(v_0)\|_{L(H_{1-\beta} \to H_\beta)} \le c_1, \quad c_1 \in \mathbf{R}^+ . \tag{7.91}$$

We shall use the representation of problem (7.1) in the form (7.4) and $X_{0,\alpha}$ as the phase-space, supposing that $\alpha$ is subject to the restrictions (7.6).

The next inequality follows from (7.91):

$$\sup_{v_0 \in \mathcal{P}(\alpha)\mathcal{A}} \|f'(v_0)\|_{L(H_{1-\beta,\alpha} \to H_{\beta,\alpha})} \le c_1(\alpha) , \tag{7.92}$$

where $\mathcal{P}(\alpha)$ is the orthoprojector of $X_{0,\alpha}$ onto $H_{1,\alpha}$.

Let us denote by $V_t(\alpha): X_{0,\alpha} \to X_{0,\alpha}$ the solution operators of problem (7.4). It is easy to verify that they have the same properties as those $V_t: X_0 \to X_0$ of problem (7.3). Denote by $U(t, \alpha, \vec{\Psi})$ the differential of $V_t(\alpha)$ at the point $\vec{\Psi} \in \mathcal{A}$. The family $\{U(t, \alpha, \vec{\Psi}): X_{0,\alpha} \to X_{0,\alpha}, t \in R^+\}$ is a collection of solution operators of the linear problem

$$\partial_t \vec{u}(t,\alpha) = a(\alpha)\vec{u}(t,\alpha) + B(t,\alpha,\vec{\Psi})\vec{u}(t,\alpha)$$
$$\equiv L(t,\alpha,\vec{\Psi})\vec{u}(t,\alpha)\,, \tag{7.93}$$
$$\vec{u}|_{t=0} = \vec{\xi}\,,$$

where $B(t,\alpha,\vec{\Psi})\vec{u}(t,\alpha) = \left(\begin{smallmatrix} 0 \\ -f'(v_0(t,\alpha,\vec{\Psi}))u_0(t,\alpha) \end{smallmatrix}\right)$ (see (7.61)).

According to Theorems 4.8 and 4.9 we have to majorize the quadratic form $(L(t,\alpha,\vec{\Psi})\vec{u},\vec{u})_{X_{0,\alpha}}$ for an arbitrary $\vec{u} \in X_{0,\alpha}$. From (7.10) and (7.91), we get (as in (7.16)):

$$
\begin{aligned}
((L(t,\alpha,\vec{\Psi})\vec{u},\vec{u})_{X_{0,\alpha}} \\
&\leq -\alpha\|u_0\|_{1,\alpha}^2 - (\nu-\alpha)\|u_1\|^2 \\
&\quad - (f'(v_0(t,\alpha,\vec{\Psi}))u_0, u_1) \\
&\leq -\alpha\|\vec{u}\|_{X_{0,\alpha}}^2 + \|f'(v_0(t,\alpha,\vec{\Psi}))u_0\|_{\beta,\alpha}\|u_1\|_{-\beta,\alpha} \\
&\leq -\alpha\|\vec{u}\|_{X_{0,\alpha}}^2 + c_1(\alpha)\|u_0\|_{1-\beta,\alpha}\|u_1\|_{-\beta,\alpha} \\
&\leq -\alpha\|\vec{u}\|_{X_{0,\alpha}}^2 + c_2(\alpha)\|\vec{u}\|_{X_{-\beta,\alpha}}^2\,,
\end{aligned} \tag{7.94}
$$

where $c_2(\alpha) = \frac{1}{2}c_1(\alpha)$. Thus, the condition (4.34) of Theorem 4.8 is fulfilled with $h_0(t) = \alpha$, $m = 1$ and $h_{s_1}(t) = c_2(\alpha)$, $s_1 = -\beta$. Therefore $\dim_H(\mathcal{A}) \leq N$ where $N$ is the minimal integer satisfying the inequality

$$-\alpha N + c_2(\alpha)\operatorname{Sp}_N \vec{A}^{-\beta}(\alpha) < 0\,. \tag{7.95}$$

Let us remember that

$$\vec{A}(\alpha) := \begin{pmatrix} A(\alpha) & 0 \\ 0 & A(\alpha) \end{pmatrix} \qquad A(\alpha) = A - \alpha(\nu-\alpha)I\,.$$

The spectrum of $\vec{A}(\alpha)$ consists of the numbers $\lambda_k^{\pm}(\alpha) = \lambda_k(\alpha) = \lambda_k - \alpha(\nu-\alpha)$, $k = 1,2,\ldots$ where $0 < \lambda_1 \leq \lambda_2 \leq \ldots$ are eigenvalues of $A$.

In order to estimate $\dim_f(\mathcal{A})$ we suppose that

$$\operatorname{Sp}_n \vec{A}^{-\beta}(\alpha) \leq c_3(\alpha)n^{1-\beta}, \quad n = 1,2,\ldots \tag{7.96}$$

with some $c_3(\alpha) \in \mathbf{R}^+$. Then, according to Theorem 4.9 $\dim_f(\mathcal{A}) \leq N$, where $N$ is the minimal integer such that

$$-\alpha + c_2(\alpha)c_3(\alpha)N^{-\beta} < 0\,, \tag{7.97}$$

We know that $X_{0,\alpha}$ and $X_0$ coincide as sets and their norms are equivalent. Therefore the dimensions $\dim_H(\cdot)$ and $\dim_f(\cdot)$ of $\mathcal{A}$ as subsets of the space $X_{0,\alpha}$ and of the space $X_0$ are the same. So we have proved the following theorem:

### Theorem 7.9

*Let the conditions of Theorem 7.5 and condition (7.91) be fulfilled, and $\alpha$ be a number satisfying the inequalities (7.6). Then $\dim_H(\mathcal{A}) \leq N$ where $N$ is the minimal integer satisfying the inequality (7.95). If the spectrum $\{\lambda_k\}_{k=1}^{\infty}$ of operator $A$ is such that the inequality (7.96) is satisfied, then $\dim_f(\mathcal{A}) \leq N$ where $N$ is the minimal integer satisfying (7.97).*

# References

[1] Ladyzhenskaya, O. On the dynamical system generated by the Navier–Stokes equations. *Zapiskii of nauchnich seminarovs LOMI. Leningrad.* **27** (1972) 91–114 (English translation in *J. of Soviet Math.* **3** 4 (1975))

[2] Mallet-Paret, J. Negatively invariant sets of compact maps and an extension of a theorem of Cartwright. *J. Diff. Eq.s* **22** (1976) 331–48

[3] Douady A., & Oesterlé J. Dimension de Hausdorff des attracteurs. *C. R. Acad. Sc.* **290** (1980) 1135–38

[4] Foias, C., & Temam, R. Some analytic and geometric properties of the solutions of the Navier–Stokes equations. *J. Math. pures and appl.* **58**, fasc.3 (1979) 339–68

[5] Il'ashenko, U. S. Weak contractive systems and attractors for Galerkin's approximations of the Navier–Stokes equations. *Uspechi Math. Nauk.* **36**, 3 (1981) 243–4

[6] Ladyzhenskaya, O. On finite dimensionality of bounded invariant sets for the Navier–Stokes equations and some other dissipative systems. *Zapiskii nauchnich seminarovs LOMI* **115** (1982) 137–155

[7] Babin A. V., Vishik M. I. Attractors for the Navier–Stokes system and parabolic equations and an estimation of its dimension. *Zapiskii nauchnich seminarovs LOMI* **115** (1982) 3–15

[8] Il'ashenko, U. S. On the dimension of attractors for $k$-contract-

ive systems in an infinite-dimensional space. *Vestnik MGU. ser. math. and mech.* **3** (1983)

[9] Constantin P., Foias C., & Temam R. Attractors representing turbulent flows. *Preprint Orsay, France* **35**, n. 695 (1984) 1–67

[10] Ghidaglia I. M., & Temam R. Attractors for Damped Nonlinear Hyperbolic Equations. *Preprint Orsay, France* **39**, n. 792 (1984) 1–71

[11] Ladyzhenskaya, O. On estimates of the fractal dimension and the number of determining modes for invariant sets of dynamical systems. *Zapiskii nauchnich seminarovs LOMI* **163** (1987) 105–29

[12] Ladyzhenskaya, O. On finding the minimal global attractors for the Navier–Stokes equations and other PDE. *Uspechi Math. Nauk.* **42**, n. 6 (1987) 25–60

[13] Hale, J. K. *Theory of functional-differential equations. Springer-verlag, Berlin–Heidelberg–New York,* (1977)

[14] Henry, D. *Geometric theory of semi-linear parabolic equations. Springer-Verlag, Berlin–Heidelberg–New York,* (1981)

[15] Ladyzhenskaya O., Solonnikov V., & Uralcéva N. *Linear and quasilinear equations of parabolic type. M., Nauka* (1967)

[16] Ladyzhenskaya O., & Uralcéva N. Survey of results on solvability of boundary value problems for uniformly elliptic and parabolic quasilinear equations of second order with unbounded singularities. *Uspechi Math. Nauk.* **41**, n. 5 (1986) 59–83

# Index